Genetic E

Dedicated to my parents, Patrick and Angela Farrelly

Genetic Ethics
An Introduction

Colin Farrelly

polity

First published in 2018 by Polity Press

Polity Press
65 Bridge Street
Cambridge CB2 1UR, UK

Polity Press
101 Station Landing
Suite 300
Medford, MA 02155, USA

ISBN-13: 978-0-7456-9503-7
ISBN-13: 978-0-7456-9504-4 (pb)

A catalogue record for this book is available from the British Library.

Library of Congress Cataloging-in-Publication Data
Names: Farrelly, Colin Patrick, author.
Title: Genetic ethics : an introduction / Colin Farrelly.
Description: Cambridge, UK ; Medford, MA, USA : Polity Press, [2018] |
    Includes bibliographical references and index.
Identifiers: LCCN 2018007933 (print) | LCCN 2018023953 (ebook) | ISBN
    9780745695075 (Epub) | ISBN 9780745695037 | ISBN 9780745695044 (pb)
Subjects: LCSH: Human genetics--Moral and ethical aspects. | Genetic
    engineering--Moral and ethical aspects.
Classification: LCC QH438.7 (ebook) | LCC QH438.7 .F37 2018 (print) | DDC
    174.2/96042--dc23
LC record available at https://lccn.loc.gov/2018007933

Typeset in 10.5 on 13 pt Plantin by
Servis Filmsetting Ltd, Stockport, Cheshire
Printed and bound in the United Kingdom by Clays Ltd, Elcograf S.p.A

For further information on Polity, visit our website: politybooks.com

Contents

Introduction

I Genetic engineering: tomorrow's reality?

Advances in the biomedical sciences, especially our understanding of the role genes play in the development of different *phenotypes* – that is, observable characteristics such as health, disease and behaviour – might help us advance important moral aspirations. Those aspirations range from preventing and treating specific diseases to realizing greater equality of opportunity, the healthy aging of a population, and expanding the scope of reproductive freedom.

The Constitution of the World Health Organization defines "health" as follows: "Health is a state of complete physical, mental and social well-being and not merely the absence of disease or infirmity."[1] The genes we are born with can have a very significant impact on our physical, mental and social wellbeing. And, as such, I believe a book dedicated to an exploration of genetics and ethics is warranted. A better understanding of our biology has led to the advancement of new technologies such as genetic screening and testing, gene therapy and genome editing (also called gene editing). Worldwide approximately 2,500 gene therapy clinical trials have already taken place.[2] With gene therapy, genetic material is inserted into the cells to correct the effect of a mutated gene that is causing disease. So gene therapy could replace a mutated gene that is causing disease with a healthy copy of the gene or make the mutated gene that is functioning improperly "inactive." Genome editing is a much more recent, and precise, technique of genetic manipulation. It involves adding, removing or altering DNA in the genome. CRISPR-Cas9 is a technique that offers scientists a more simple method of manipulating genes, and CRISPR was approved for human clinical trials in the United States in 2016. And in the summer of 2017 *Nature* published a landmark study (Ma et al. 2017) on gene editing of human embryos (embryos not destined for implantation).[3] Gene editing was utilized to target a mutation in a gene called MYBPC3, a gene that causes the heart muscle to thicken. Such thickening can lead to a fatal heart condition in young athletes. This early experiment of gene editing in human embryos raises the prospect of being able to correct potentially deleterious germline mutations before a child is even born.

Advances in our knowledge of genetics might go much further than simply allowing us to treat and prevent disease via gene therapy or genome editing. An understanding of the role genes play in human intelligence, memory, emotional resilience, moral behaviour and happiness could expand the domain of interventions (both environmental and genetic) at our disposal to improve our opportunities for living flourishing lives. Advances in screening human embryos prior to implantation (known as pre-implantation genetic diagnosis, or PGD) for sex or a higher propensity towards desirable behavioural characteristics could expand, for better or worse, the discretionary power parents have in shaping the potential identify of their offspring.

Is the prospect of "genetic engineering" humans something to be hailed or feared? Will it improve our prospects for living flourishing lives and creating more fair societies or will it threaten our very survival and/or create greater inequality and other grave societal problems? These are among some of the most pressing, and novel, societal concerns of the twenty-first century. And these are questions we shall address, and attempt to answer (even if only provisionally), in the chapters to follow.

II Beneficence or precaution?

One could explore the ethical and societal implications of advances in our understanding of genetics from many different moral lenses and principles. For example, suppose one started from a duty to aid (or principle of beneficence). Peter Singer, a utilitarian and arguably the most influential living philosopher in applied ethics, invoked what he called the *principle of preventing bad occurrences* to raise greater awareness about the problem of global poverty. This principle states:

> If it is in our power to prevent something bad from happening, without thereby sacrificing anything of comparable moral importance, we ought, morally, to do it. (Singer 1972: 231)

Singer asked us to contemplate a now famous thought experiment to demonstrate the normative force of this duty to aid. The example concerns a child who is drowning in a shallow pond. You are walking past the pond and notice the distressed child in need of assistance. The child is not your child, nor a compatriot, but a citizen from a distant and far-away country. Nevertheless, the child is *a human being* in need of assistance. If the only burden to be incurred by saving the child is getting one's shoes and trousers wet, then there is, argues Singer, a stringent duty to do so.

Singer then drew an analogy between the example of the drowning child and global poverty. The rich living in the developed world have a stringent moral obligation, he argued, to donate a significant amount of their income to help those living in poverty in distant lands. Singer's argument spurred much debate on the demands of *global justice*, a topic largely ignored by philosophers before the publication of Singer's article. Questions such as "Do national boundaries have any ethical significance?" are still debated over forty years later.

Despite its potential usefulness, there are also significant limitations in invoking Singer's principle of preventing bad occurrences. Most of the bad things in the world, including global poverty, are infinitely more complex and complicated than the example of helping a drowning child in a shallow pond. How do we ensure the actions we undertake to redress poverty *actually help* others[4] rather than just wasting our time and energy or, even worse, making the situation even more dire (as can conceivably happen in the case of providing foreign aid)?

The problem of global poverty is not simply, or even primarily, a problem of the rich not donating money to the poor. But it is hard not to form that impression from Singer's original article and moral argument. The central moves in his moral argument are (1) to invoke the principle of bad occurrences, then (2) to link that principle with the badness of poverty, and then (3) to conclude that the solution to this bad is for the rich to donate more money to foreign aid.

Suppose we ran a similar moral analysis to buttress the case for mitigating the genetic lottery of life. Imagine the child in need of assistance was not drowning in a shallow pond. Instead, the source of the threat of the child drowning was *internal* to her, as she was born with cystic fibrosis (CF), a genetic disorder which impedes the normal functioning of lungs. Left untreated, a child will in effect "drown from within" as the condition fills her lungs with fluid.

In the 1970s a child with CF had a very low life expectancy at birth – typically not more than just a few years. However, by the 1990s things had improved. The median age of survival for a child born with CF in Canada was nearly thirty-two years, and that had increased to nearly fifty years by 2012 (Stephenson et al. 2015). However, a life expectancy of fifty years is still thirty years less than the average in Canada. Inheriting the genes for CF has a profound impact on the life prospects of a person.

Dying from CF, like dying from poverty, is a bad thing we should seek to prevent if possible. So why not make the Singerian moves (1) invoke the principle of preventing bad occurrences, then a modified version of (2), to focus on the harms of genetic disease, and then conclude that (3) people should be donating all their surplus resources to creating a gene therapy for CF, until the sacrifice risks something of comparable importance to developing the disease?

One of the central limitations of invoking the principle of preventing bad occurrences and applying it to one specific form of badness (be it poverty or CF) is that the world has many bad things about it that need to be addressed. So invoking the principle itself doesn't help us determine how to prioritize among the plurality of problems (i.e. bad occurrences) we need to address, nor does it bring adequate attention to the realities that different kinds of intervention will be more risky, or costly, or effective than other forms of intervention.

Rather than invoking a duty to aid in order to support the prospect of intervening in our genes, one could adopt a very different moral perspective. Some might be inclined to urge *caution* rather than optimism. Over the past number of decades many scholars and activists, especially environmentalists, have invoked what is called the "precautionary principle" to oppose what they see as potentially harmful technological advancements, for example genetically modified crops.

With most technological advancements (e.g. industrialization, capitalism, medicine, computers, etc.) we often do not know with certainty *all* the potential benefits and harms that could result from the application of the technology. And this often leads critics to urge caution, even to the extent of not pursuing the technology given the uncertainty concerning its risks. When we lack full knowledge, as is often the case with most decisions in life, how risk averse should we be?

It can be challenging to define the precautionary principle because it is often invoked and defined in different ways. VanderZwaag (1999), for example, identifies fourteen different versions in various treaties and declarations. At the most basic level, the idea of the precautionary principle embodies the catchphrase "better safe than sorry" (Sunstein 2003: 1004). A critic of genetic interventions, for example, might invoke the "better safe than sorry" sentiment to oppose genome editing, increasing the human lifespan or permitting sex selection. Unless we know with certainty that such interventions will not cause unintended harms – such as adverse side-effects in the case of genome editing or gene therapy, overpopulation and/or an increase in climate change as a result of increasing the lifespan, or a distorted sex ratio because of sex selection techniques – then such technologies should not be permitted.

I believe a wise society would of course aspire to protect health, the environment and a balanced sex ratio, but I do not believe it would embrace a stringent precautionary principle to oppose the development of new genetic interventions. Consider, for example, one popular version of the precautionary principle often invoked (known as the Wingspread Declaration):

> When an activity raises threats of harm to human health or the environment, precautionary measures should be taken even if some cause-and-effect relationships are not established scientifically. In this context the proponent of the activity, rather than the public, should bear the burden of proof. (Sunstein 2003: 1006)

Suppose one worries that the development of an aging intervention, for example, will lead to problems of population density and more climate change. One might not know with absolute certainty that intervening in the aging process will cause these problems, just that there is *some* (maybe even low) risk, and that it is "better to be safe than sorry." Invoking the precautionary principle places the burden on those arguing in favour of intervening in the aging process to establish that these problems will not arise, and if that burden of proof cannot be met then the goal of retarding aging should not be pursued. Is this a reasonable way to approach the regulation of new genetic interventions, or any technological advancement? Would it be sensible to invoke these same worries for other public health measures that could increase life expectancy, such as exercise, smoking cessation, or higher compliance with prescribed medications that help manage specific chronic conditions?

Any application of a strict application of the precautionary principle is rife with problems. As the legal scholar Cass Sunstein has argued, "any effort to be universally precautionary will be paralyzing, forbidding every imaginable step, including no step at all" (Sunstein 2003: 1008). When a critic of any genetic intervention, for example an aging intervention, raises the concerns that such an intervention has risks, they are focusing only on the risks of intervention. But a sage society must weigh these against the risks of inaction. Sunstein notes that people are often susceptible to the "Myth of a Benevolent Nature," the "mistaken belief that nature is essentially benign, leading people to think that safety and health are generally at risk only or mostly as a result of human intervention. A belief in the relative safety of nature and the relative risk of new technologies often informs the precautionary principle" (ibid.: 1009).

The critic who opposes an aging intervention by noting the potential risks of such a technology ignores the risks, to human health and economic prosperity, of *not* intervening in the aging process – the dramatic rise in chronic diseases for aging populations, growing care requirements to attend to those aging populations, and the impact an aging population will have on the economic prosperity of a society if the ratio of workers to retired persons alters dramatically from historical norms. We know, for example, with great scientific certainty, that normal aging profoundly impacts a person's health in late life. Aging eventually causes multi-morbidity for most humans. The end of life can be a painful, and expensive, period of chronic disease and disability. A virtuous society (or "polity," the term I prefer to use) would give attention to the risks and harms of aging, as well as the potential harms of altering the aging process. Selectively applying the precautionary principle to oppose intervening in aging contravenes the "intellectual virtues" of having the ability to recognize the salient facts and a sensitivity to details. When a principle such as the precautionary principle is invoked to bring attention only to one set of risks, it skews the exercise of intellectual virtue.

Some caution is definitely prudent when it comes to the prospect of genetically engineering humans. How could we ethically experiment on humans for something such as gene therapy, an intervention that involves inserting genes into a patient's cells? To assess its safety and efficacy we should approach gene therapy or editing experiments with the same rigorous standards we enforce for other medical interventions rather than simply invoking an emotive and dismissive stance. In the procedure known as angioplasty, for example, stents are used to inflate balloons which stretch blocked arteries. As we age, it is common for our coronary arteries to harden, and this can cause coronary heart disease. No one (at least to my knowledge!) protests medical advances such as the use of stents on the grounds that we are "playing God!" or that inserting tiny balloons in our arteries is "unnatural." Furthermore, there are risks and costs associated with angioplasty, the most common of which are blood clots and bleeding. But there are also rare and more serious risks, such as heart attack and stroke. And, as a surgical procedure, angioplasty has non-trivial costs. In a country such as the United States it would cost thousands of dollars to undergo this procedure.

I do not mean to imply that the principles of beneficence and precaution have no legitimate role to play in our moral deliberations. Rather, my worry is that, when we frame the prospect of "genetic intervention" in the narrow fashion that is typically invoked by such principles, it impedes our ability to adopt a "bird's eye perspective" of the moral landscape.

III Hitting the "re-set" button on debates about genetic engineering

When I decided to undertake writing *Genetic Ethics*, I spent the first year contemplating the following question: How can educators help stimulate greater interest in, as well as rational and cogent thought and argument about, the *ethical significance* of advances in human genetics? This pedagogical question is the *foundational* question that orients the normative analyses and arguments developed in this book.

While researching and writing about these contentious ethical topics for nearly two decades now, I have discovered that it is not easy to answer this pedagogical question. I often found myself disappointed by some of the argumentation developed by prominent political theorists and bioethicists who have offered normative insights into the topic of genetic intervention. In his short book titled *The Case against Perfection: Ethics in the Age of Genetic Engineering*, Sandel argues that a quest to perfect our biology "threatens to banish our appreciation of life as a gift, and to leave us with nothing to affirm or behold outside of our own will" (Sandel 2007: 99–100). And Habermas argues that parents selecting the genetic constitutions of their offspring threaten the self-understanding of the affected person as an autonomous and responsible agent. Eugenic practices, he argues, "might well harm the status of the future person as a member of the universe of moral beings" (Habermas 2003: 79).

I do not think that such contributions to the debates on the ethics of genetic engineering have helped to stimulate informed, rational and cogent ethical deliberations. Instead, I believe they have helped confuse and muddle the dialogue. Should we not undertake scientific research that might permit us to improve our health, behaviour, intelligence and happiness because doing so could "leave us with nothing to behold outside of our own will" or because "it might harm the status of future persons as members of the universe of moral beings"?

I do not believe such a high level of philosophical dialogue, cast over many distinct potential technological innovations (labelled generally as "eugenics" or "genetic engineering"), is very helpful. In fact I think such arguments have obscured, rather than helped clarify, the distinct moral stakes involved in moving forward, or remaining stagnant, with respect to developing novel genetic interventions. Rather than approaching the issue of genetic engineering from the conviction that what is most important is that we must appreciate life as a gift or protect the status of future persons as members of the universe of moral beings, I believe a much better moral orientation is one that prescribes that we, as a society, aspire to exemplify moral and epistemic (or intellectual) *virtue* (rather than *vice*). Such a moral orientation, we shall soon see, does not lead to broad conclusions about the desirability of, or problems inherent with, genetic engineering. It inspires a curious and optimistic, yet also cautious and provisional, attitude towards the different ways we could improve the opportunities for humans to live flourishing lives.

And rather than develop an ethical analysis that invokes abstract hypothetical examples that presuppose new genetic interventions of certain sorts will soon exist, I have gone to great lengths to focus the ethical analysis on where the science is today, with the diverse complexity of issues that arise at the ground level of the basic scientific research and public policy. These range from concerns about what medicine is for, and the safety, efficacy and costs of genetic intervention, to the prospects of genetic discrimination and concerns about the potential harmful impacts of increasing the human lifespan and permitting sex selection.

There is another current of the debates about the ethics of genetic intervention that I also think has been unhelpful. And that concerns the invocation of "normal species functioning" (Buchanan et al. 2000; Daniels 2000) and the therapy/enhancement distinction (Resnick 2000). Therapeutic interventions are those that seek to prevent or treat a disease, such as gene therapy for cystic fibrosis or Parkinson's disease. Norman Daniels (1985) has argued that justice requires restoring "normal species functioning," so, applied to genetics, this means there is a strong presumption in favour of the idea of a genetic decent minimum (Buchanan et al. 2000). And, while he does not rule out the prospect that some "enhancements" could be considered as required by justice, Daniels maintains that public policy is served well by retaining the therapy/enhancement distinction and that justice might require restricting access to some enhancements.

I will not provide an exhaustive critique here of what I think is problematic with the idea of "normal species functioning" and the therapy/enhancement distinction. I believe such a framing of the issues skews our moral deliberations because it provides an illusory "natural" baseline – a baseline people are prone to want to aspire to maintain and preserve because they think it is either "natural" or "the best." Utilizing the idea of "normal species functioning" or what is "natural" with respect to our genetic endowments is misguided because, as is noted in the National Academies of Sciences, Engineering, and Medicine's recent report on the ethics and governance of genome editing, "there is no single 'normal' human genome sequence; rather, there are multiple variant human genomic sequences ..., all of which occur in the worldwide human gene pool and, in that sense, are 'natural,' and all of which can be either advantageous or disadvantageous" (NASEM 2017: 106).

Genetic mutations can be *both* adaptive *and* maladaptive, depending on the environment in which we live. Consider, for example, the genetic underpinnings of a trait such as aggression (Smith and Harper 1988). "A detailed evaluation and meta-analysis of 24 genetically informative studies concerning aggression concluded that heritability accounted overall for about 50% of the variance" (Craig and Halton 2009: 102).

One type of aggression, typically associated with psychopathy, arises from a lack of emotional sensitivity. "Psychopaths appear to lack emotional distress and are impervious to distress in others" (Glenn et al. 2011: 372). The neurobiology of a psychopath's brain is different than that of the average person. All people can become angry, indeed even potentially violent (though men much more so than women). Most people can regulate such emotions as anger in ways that prevent them from engaging in problematic behaviours that erode their social relationships and are self-destructive (maladaptive behaviour). But a psychopath's brain is prone to engage in predatory behaviours. They are very manipulative, have a sense of grandiosity and are egocentric. Psychopathy is a personality disorder; the behaviours of a psychopath are *maladaptive*.

But why does psychopathy or, more precisely, the range of personality traits that fall under the broad category of the disorder exist as a range of possible behaviours for humans? What is the ultimate or evolutionary causation of such maladaptive behaviour? The ultimate causes of maladaptation are aspects of genetic systems in relation to changes in environments (Crespi 2000: 624). Different theories have been offered to explain why psychopathy might actually be adaptive rather than pathology.[5] One theory is that "psychopathic traits have been selected for because they offer a fitness advantage in specific environments" (Glenn et al. 2011: 371). Having little remorse or empathy might prove advantageous to individuals struggling for survival and competing for mates in the hostile environment typical of our evolutionary past. The personality traits of a lack of remorse, guilt and empathy for others can actually be adaptive in certain environments. But these same behaviours can be erosive of the social skills needed to flourish in the modern world, where emotional sensitivity helps one succeed with interpersonal relationships (e.g. marriage, work, parenting, friendships, etc.).

A second example that shows that particular genes that we now consider maladaptive might have, historically, been adaptive are the genes implicated in obesity.[6] There are monogenetic and polygenetic forms of obesity (Choquet and Meyre 2011). Genes associated with increased feelings of hunger, increased snacking and decreased satiety are maladaptive in an environmental context such as contemporary America, where cheap, high calorie food is easily available. But in our evolutionary history food was often scarce, and so these same genetic mutations could have been adaptive in those environmental circumstances where the risks of malnourishment and starvation were much higher than the health risks of obesity. In short, the accumulation of fat tissue that causes obesity today provided a fitness advantage in our evolutionary history, a history plagued by famine.

Our ability to regulate food intake via genetic mutations that help us store body fat could be beneficial if we lived in hunter-gatherer societies where food was typically scarce. But in the environment of a developed country such as the United States, where cheap, high calorie food and sugary drinks are in abundance, possessing such genes could pose a serious health threat.

As the examples of aggression and obesity make clear, the idea of utilizing "normal species functioning" or what is "natural" with respect to our genetic endowments as some normative guideline is misguided. The "it's unnatural!" objection to genetic engineering humans is folly because it treats the biology humans happen to have at this moment in time in our evolutionary history as "special," as something to be protected and maintained. But the reality is that our biology is constantly evolving and changing. When there is variation in traits, differential reproduction and heredity, then in time more advantageous traits (with respect to *reproductive fitness*) become more common in the population. This is the process of *evolution by natural selection*. And the genetic endowments of the population today, in the twenty-first century, are the result of evolution by natural selection over thousands of years.

These genetic endowments will continue to evolve and change even if we do not pursue genetic engineering. The question we thus face is whether it is preferable to let that blind and arbitrary process shape our biology (thus impacting our life prospects, for better or for worse) or if it is preferable to permit human intention and innovation to shape that biology as well, by consciously and purposively altering genes (for example via gene therapy or genome editing) and/or gene expressions (by modulating environment or behaviour).

A helpful way to begin to think about the ethics of "genetic engineering" is to draw an analogy with the ethics of *social engineering*. Culture is social engineering. Different institutions and cultural practices – from the family, political economy, religion, constitution, patriotism and education – help to contribute, for better or for worse, to engineering a certain type of environment for humans. Is social engineering beneficial or detrimental? The answer is "yes," it can be both beneficial and detrimental.

Social engineering is not necessarily good or bad. The family, for example, can be an immensely beneficial institution for individuals and societies when it helps foster feelings and sentiments of love and commitment between intimates, helps motivate and sustain planning a life together (including financial security), and encourages a nurturing and safe environment for having and raising offspring.

But the family can also be oppressive and unjust. For example, in a culture where women are economically dependent upon men – when marriage to a man is an *economic necessity* for a woman – it is a harmful form of social engineering. It undermines equality between the sexes. The same is true of a culture where young girls are raised with the expectation that becoming a mother is the only aspiration they should strive to fulfill, or when women are expected to perform the lion's share of unpaid domestic labour within the home, or when it is only mothers (but not fathers) who are expected to "recalibrate" their career aspirations when they become a parent.

A family modelled after an ideal of equal partnership can be a desirable form of social engineering, whereas a model of the family predicated upon patriarchal relationships is not. The same can be said for the political economy of a country. Economic policy is a form of social engineering; humans have experimented with free market economies, command economies and the welfare state. The economy can be induced, via economic policies, to expand or compress, to distribute wealth and income more or less equitably, etc. A political economy where economic development stagnates and creates massive unemployment is not desirable, whereas an economic intervention that grows the economic prosperity of everyone, especially the poor, is beneficial and more desirable from the standpoint of distributive justice.

A sense of patriotism, instilled through public education, the media, sports, the national anthem, etc., can help foster a sense of solidarity and democratic engagement within a culturally diverse polity. But rampant nationalism can lead to intergroup conflict and cultural intolerance. All forms of social engineering could be used for good or bad purposes. My point is, we should not generalize and say social engineering is intrinsically desirable and genetic engineering is intrinsically undesirable. The devil is really in the details. Both forms of engineering could be used for good or bad purposes.[7] It depends upon (1) the aim or aspiration behind social and/or genetic engineering, (2) how successful we can expect to be in terms of realizing those expected values and aims, and (3) what the potential cons or drawbacks of such engineering might be (e.g. unintended consequences).

Each specific form of engineering, be it social or genetic, must be considered on its own merits and demerits rather than treated as one *monolithic* type of intervention, social or genetic. This is something that will become apparent, I hope, in the virtue ethics analysis developed to assess different types of genetic engineering.

The normative analysis advanced in this book is not deferential to "normal species functioning" or the therapy/enhancement distinction. I believe any potential genetic intervention – in the aging process, the psychological immune system, or intelligence and behaviour – should be considered on its own merits and demerits.

An intervention that alters the aging process in humans, for example, can be considered an "enhancement" in that it might lead to an increase in the average human lifespan. But it is also *preventative medicine* in the sense that it would reduce our risks of morbidity, disability and death. The philosopher John Harris nicely captures the deficiency of utilizing "normal species functioning" as some moral baseline in our theorizing about the ethics of genetic intervention:

> To define disease as deviations from normal species functioning means that treating the diseases of old age would not be therapeutic (indeed, the diseases of old age would not be diseases) in any sense because diseases of old age are species typical (or of course constitute normal species functioning), and it is species typical and a part of normal functioning that we cease to function in old age and that we die. (Harris 2007: 45)

Many of the most important public health measures, from immunizations and exercise to the sanitation revolution, can be classified as "enhancements," and I think they can be much more significant than some potential "therapeutic" interventions that have only a marginal impact in terms of helping people manage the multi-morbidities of late life. Once again I will approvingly quote John Harris when he claims that "the overwhelming moral imperative for both therapy and enhancement is to prevent harm and confer benefit. Bathed in that moral light, it is unimportant whether the protection or benefit conferred is classified as enhancement or improvement, protection or therapy" (Harris 2007: 140).

In this book I want to hit the "reset" button on the ethical debates about genetic engineering. To help shift the dialogue and debate in a direction quite different from Sandel and Habermas, as well as away from the fixation on the therapy/enhancement distinction, I will not be addressing (beyond the brief comments here) any of those arguments further. I want to keep the focus of the normative analysis that follows on the issues I believe will have greater pedagogical value for students and scholars in the life sciences and humanities/social sciences.

IV What is virtue ethics?

What does virtue ethics maintain? When thinking about what actions we should pursue, or beliefs we should form, either as individuals or collectively as a polity, virtue ethics prescribes we should mimic the actions and beliefs that a virtuous agent would pursue and adopt if they were in our circumstances. Resnick summarizes virtue ethics succinctly as follows:

> Virtues are character traits (or habits) that we develop over time, through practice and imitation. We learn how to practice virtue by following the example of other people who exhibit virtue in their conduct and by applying this example to our own conduct. In this way, virtue is similar to other practical skills, such as carpentry, driving, or piano playing. All virtues are means between extremes of behavior. (Resnick 2012: 334)

In Plato's *Republic*, the character of Socrates reveals the limitations of conceiving of morality or justice in simple rule-like terms. A popular moral conviction of ancient Athenians was that justice requires telling the truth and returning anything borrowed. Socrates reveals the shortcomings of this rigid conception of justice by asking us to consider the following kind of case. Imagine you borrow a weapon from a neighbour, so you can go hunting that afternoon. Later that evening you return home and hear raised voices and a heated argument coming from your neighbour's house. The neighbour storms over to your house, banging on the door, demanding the return of his weapon. You know he is in an emotionally distressed state, and his reason for wanting the weapon returned at that exact moment is he intends to harm someone.

In such circumstances, does justice require you to tell the truth and return the weapon, thus facilitating the harm, perhaps even death, of an innocent third party? Obviously not. A strict application of the rule "Always tell the truth" would not be moral. A virtuous agent would recognize that telling a "white lie" in such circumstances is the appropriate action. You tell the neighbour you forgot the weapon at a friend's house and will collect it the next day. This will give the neighbour time to calm down, so that harms committed in the heat of the moment can be avoided.

For the ancient Greek philosophers, *character* (rather than rules or principles) was the central focus of moral evaluation. Plato's star pupil, Aristotle, developed the most influential account of virtue ethics in *Nicomachean Ethics*. Aristotle argued that the virtuous agent possesses the right amount of emotion – the "mean" between the extremes of having an excess of the feeling on the one hand and a deficiency of the emotion on the other hand. The courageous soldier or polity, for example, is neither a coward (deficiency of the emotion) nor too rash (excess of the emotion). Courage is displayed when you stand your ground and fight when doing so is wise and appropriate. But always cowering away from any and all challenges or, conversely, being so brash as to take on all potential enemies for even trivial reasons are examples of *vice* rather than virtue.

This idea that the virtuous agent occupies the mean between extremes is helpful in framing the overall attitude we should have with respect to the prospect of genetic interventions. Should we be fearful about such new technologies – that intervening in our genes is playing God, will repeat the eugenic atrocities of the past, or lead to greater inequalities between the "haves" of this world and the "have nots"? If fear is the only emotion that is elicited, then we may be persuaded to eschew going down the path of purposefully altering our biology with new biomedical technologies.

Alternatively, if cock-eyed optimism is our mindset, we might take the view that health innovations that utilize the latest findings in genetics should be the top priority for addressing the pressing societal problems of the twenty-first century. Indeed, this zealous attitude could be taken so far that we naively think that genetic interventions will make all conventional medical or preventative health measures obsolete.

We can think of these two positions as opposite ends of the spectrum for *excessive fear* and *excessive enthusiasm*, represented by what I will call the Biomedical Luddite Society and the Biomedical Technophile Society:[8]

> *Biomedical Luddite Society*: Eschews the acquisition of new knowledge about genetics and any prospect of genetic intervention.

> *Biomedical Technophile Society*: Adopts the attitude that advances in our understanding of genetics and potential genetic interventions will remedy the most significant societal problems likely to arise for the foreseeable future.

The biomedical Luddite attitude might be predicted upon (a) the assumption that the harms of genetic intervention will far outweigh any benefits (genetic *precautionary principle*); and/or (b) a belief that conventional knowledge and medical interventions (e.g. more exercise, smoking cessation, better diet, chemotherapy, stents, etc.) provide more safe, effective and affordable solutions to the health challenges of today (a *status quo bias* about medical knowledge).

The Biomedical Technophile Society presupposes a faulty view of genetic determinism – that all we need to accomplish is mastery over our genetic destinies and then all of society's problems will be remedied. This overlooks the fact that the environment plays an enormous role in the development of different phenotypes (especially health). Furthermore, the Biomedical Technophile Society's limited focus on genetic interventions means it runs the risk of ignoring other laudable societal goals – such as creating a more inclusive and fair society, combating climate change and patriarchy, or eradicating poverty – goals that cannot be achieved by simply prioritizing genetic knowledge and interventions.

I believe the Luddite and technophile attitudes represent moral and epistemic *vice* rather than virtue. While I do not propose detailing all the beliefs and actions the virtuous polity would pursue (I profess that I do not know, as I am not such a virtuous agent myself!), specifying what can constitute moral and epistemic *vice* is of great service in terms of helping elucidate the folly we can be prone to. The normative analysis advanced in this book aspires to guide us away from obvious (and less obvious) vices, which I believe indirectly helps set us on the right course to the potential realization of virtue.

V The pedagogical value of virtue ethics

Two distinct features of the account of virtue ethics I employ in this book are worth noting from the start. Firstly, I am most interested in employing virtue ethics as a *public philosophy* rather than as a personal ethic. So the bulk of my focus is on collective decision-making and deliberations. The types of questions we explore include "Would a virtuous polity pursue eugenics, or aspire to retard the aging process, or permit sex selection via PGD (pre-implantation genetic diagnosis)?" Such questions are the core of the book, rather than those such as "Would a virtuous person consent to gene therapy, or aspire to slow their own rate of biological aging, or select the sex of their offspring via PGD?" The latter questions are certainly interesting and worth pondering and debating. But they are not the central focus of this book.

The virtue ethics tradition can be deployed to assess both individual actions and collective decision-making and deliberations. In *The Republic*, for example, Plato first analyzes justice at the level of society (or the *polis*) and then moves on to consider justice at the level of the individual person. The ethical analysis deployed in this book is primarily, though not exclusively, concerned with the former. I develop an account of the moral and epistemic virtues as *social* rather than personal virtues. This helps focus the ethical analysis of genetics on the broader social significance of new genetic knowledge and technologies, as well as on regulatory concerns likely to interest students and scholars in different disciplines.

This book was written and designed for students and scholars interested in the ethical and social implications of new biomedical technologies. A central goal is to integrate philosophical knowledge and an analysis of ethics with an empirical understanding of the new knowledge and innovation taking place in the biomedical sciences, with a focus on human genetics. This is not, admittedly, an easy task.

Philosophers might hope for a more nuanced and comprehensive analysis of moral and intellectual virtue than what I offer here. Philosophical debates in ethics often involve comparing and contrasting rival normative theories to see which moral framework can emerge the least unscathed after rigorous comparison and critique. But, for philosophy courses in "applied ethics" or bioethics, the tendency towards inward specialization and winning philosophical debates is typically tempered by orienting the philosophical enterprise in a more "outward looking" fashion. That is, the primary goal of the exercise is to help us make sense of the ethical and societal quandaries in which we find ourselves in the real, non-ideal, world. The primary function of such moral analyses is thus to help us figure out how to address and navigate through those quandaries rather than determine which moral theory is the best in the abstract. So I hope the approach I adopt here will appeal to philosophers looking for some "real world" focus, but I recognize it may leave other philosophy types – those wanting more theory and less of an applied focus – less satisfied.

In contrast to students in philosophy, those coming to the book from the life and medical sciences might have little interest in the philosophical nuances of ethical theories. Instead they might want to hear more details about the science and the regulatory and practical implications of research in the biomedical sciences. Again, I seek to strike a balance – the book focuses at a certain level of generality about the prospects of genetic intervention, but it does not detail, at great length, the finer points of the science involved in such innovations. It aspires to deliver something of interest both to philosophers and to those in the medical and life sciences, and the balanced "middle ground" means not too much philosophy and not too much science; the goal is to add "just enough" of each to make the project instructive to scholars in diverse fields of interest.

A good book should help provoke the reader to ponder interesting and significant questions. And my focus is more on asking questions (Is eugenics morally wrong? Should we seek to improve upon the biology we have inherited from the process of evolution by natural selection? Is it ethical to aspire to slow human aging?) than on trying to answer them definitively. The perfectly virtuous polity, or parent, or person does not exist, so asking what they would believe or do if placed in our circumstances might sound like a pointless project. But I think the primary benefit of invoking virtue ethics is that we can constructively conceptualize instances of moral and epistemic *vice*, and doing so can help clarify the moral landscape significantly for us. It can at least help us rule out certain beliefs, assumptions and courses of action because they encapsulate moral and/or intellectual *vice*.

I believe virtue ethics can be a very instructive moral lens to adopt when the issues under consideration are future (potential) technologies and innovations because it can help correct common "prospection errors" in ways that a principle-oriented lens cannot. Prospection refers to our ability to "pre-experience" the future by simulating it in our minds (Gilbert and Wilson 2005: 1352) and is a unique feature of the cognitive lives of humans. We can envision the future of our individual and collective lives in our minds. We might entertain thoughts about the likely joys and challenges of a particular career path (e.g. teacher or dentist) or decision to marry a particular person.

Social psychologists have documented a variety of cognitive limitations and biases that can arise when simulating the future in our minds. For example, such simulations can be *unrepresentative* (Gilbert and Wilson 2005: 1352). Simulations of the future are constructed from our memories. If our memory is of past eugenic policies such as sterilization of the "unfit" or of a world or society dominated by high rates of fertility and communicable disease, we might not have an accurate appreciation for the importance of aspiring to improve human biology so we can better meet the health challenges of the twenty-first century. This could skew our valuation of gene therapy or genome editing to prevent specific diseases or the importance of retarding the aging process. Historically biological aging was not a significant public health challenge for populations that experienced high rates of early and mid-life mortality. People might object to retarding aging because they believe communicable diseases kill most people alive today, and that this will also be so in the future. Or they worry that intervening in aging will cause more dire social ills, for example overpopulation. Adopting a virtue-oriented lens can help guard against the risk of making unrepresentative simulations, especially when attention is paid to the importance of the intellectual virtues (e.g. sensitivity to the salient facts).

Simulations of the future can also be *abbreviated* (Gilbert and Wilson 2005: 1353). This means that they typically focus on a few, select moments of a future event. For example, when we simulate a potential future where parents have the ability to select the sex of their offspring for non-medical purposes, through pre-implantation genetic diagnosis (meaning they screen the embryos before implantation and decide which ones to implant based on that information), we might predict it leading to a skewed sex ratio and/or exacerbating patriarchal attitudes in our society. Given the nature of the topics under discussion in this book (biomedical innovations and technologies), principle-oriented analyses of the ethics of genetic engineering of humans that champion the primacy abstract values (e.g. the primacy of reproductive freedom or precaution) can exacerbate these prospection errors. We might, as I believe Habermas and Sandel do, fixate on the potential impact such interventions might have on one value or principle (e.g. gratitude or autonomy) while ignoring other important moral values and interests. I believe a further benefit of virtue ethics, when applied to the issues of this book, is that it can help reduce the risks that our moral analyses might commit prospection errors when examining the pros and cons of potential interventions. And this is so because the virtuous agent does not fixate on one or two particular principles; instead, she adopts a wide and provisional moral lens.

VI Let's cut to the chase: what conclusions do I make?

I am very modest and provisional about coming to specific conclusions on the topics addressed in this book, in large part because I think it is too premature to estimate the extent to which advances in genetics will alter the moral landscape. Instead I aspire to help provoke the reader to see the importance of such advances to the moral aspirations of helping humans live flourishing lives. By focusing on both moral virtues (i.e. benevolence and justice) and *epistemic* virtues, I try to make the case for taking action to advance knowledge and understanding that might lead to significant health innovations (provided they are reasonably safe and cost-effective), such as gene therapy, genome editing or an aging intervention.

Let me begin by just stipulating, rather than arguing for and defending, the virtues that will be invoked in the chapters to come.

Moral virtues:
1 *Benevolence* – the virtue of benefiting others, especially preventing avoidable harms (e.g. disease or risk of disease) to both persons and populations.
2 *Justice* – the virtue of treating others fairly and impartially. This includes taking seriously the protection of basic rights and liberties and fairly distributing the benefits and burdens of social cooperation (e.g. income, healthcare resources, etc.).

Benevolence and justice can often overlap. Benevolence can move a virtuous polity to implement public policies that help benefit those in need of state assistance. A benevolent society will attend to the needs of those most at risk of disease and/or poverty, for example. The virtue of justice might prescribe many of the same courses of action or public policies. But justice adds further concerns beyond simply benefiting others. For example, justice is also concerned with ensuring those benefits are fairly distributed among the possible people who could gain advantage from particular policies or actions. Furthermore, justice will concern itself with the burdens and costs typically imposed upon some when benefiting others.

The "intellectual" or "epistemic" virtues I utilized for this book is a partial list of the virtues championed by Linda Zagzebski in *Virtues of the Mind*.

Epistemic virtues:
1 the ability to recognise the salient facts and have a sensitivity to details
2 intellectual humility
3 adaptability of intellect
4 the detective's virtues: thinking of coherent explanations of the facts.
(Zagzebski 1996: 114)

These lists of virtues are not meant to be exhaustive; instead, I think they are instructive. That is, I believe they provide a solid moral foundation for addressing the societal implications of genetics and that they can help enhance our deliberations about eugenics, gene therapy, genome editing, retarding aging, modulating cognition and behaviour and selecting the genetic traits and sex of potential offspring.

In the chapters to come I suggest that a virtue ethics lens, when applied to the issues of genetics and genetic intervention, yields a number of provisional moral theses that a virtuous polity would endorse.

1 A virtuous polity would determine if any genetic intervention, whether it be gene therapy, genome editing or a drug that activates the expression of specific genes, is morally permissible – indeed, perhaps even morally required – by its potential to prevent harm in a reasonably safe and cost-effective manner (e.g. by preventing, delaying or treating morbidity).
2 Virtuous agents would eschew both genetic determinism *and* environmental determinism.
3 A virtuous polity would not necessarily eschew eugenics, where eugenics is understood, as the philosopher Bertrand Russell defines it, as "the attempt to improve the biological character of a breed by deliberate methods adopted to that end" (Russell 1929: 254). In other words, to describe an intervention as "eugenic" does not mean it is unjust. Eugenic aspirations can be morally defensible, even morally obligatory, when they pursue empirically sound and morally justified aims (e.g. promotion of health) through reasonable and morally justified means that treat all persons as free and equal.
4 A virtuous polity would take a *purposeful* approach to determining the scope and limitations of reproductive and parental freedom. Such an approach will give due consideration to the values of autonomy, wellbeing and equality (without ascribing a primacy to any one of them).
5 A virtuous polity would aspire to promote the *healthy aging* of its population through all possible means (including interventions that extend the lifespan if doing so increased the *healthspan*). Such measures should be pursued in a responsible manner so that considerations of equity, population size, intergenerational justice and environmental impact are also taken seriously.

VII Why an introductory book on genetics and ethics?

Do advances in genetics and the biomedical sciences warrant the kind of special moral consideration accorded to them in this book? The world has so many pressing problems – ranging from war and poverty to patriarchy and climate change – that one might wonder why is it necessary to write a book focused just on issues arising from the genetic revolution. Granted a better understanding of the role genes play in different phenotypes is not going to provide a solution to all the world's problems, I do think there are some compelling reasons for engaging with the topics and issues addressed here.

The first reason stems from the fact that the genes humans inherit do often have a profound impact on our life prospects. This is most vivid at the extreme ends of the genetic lottery of life. For example, if you inherit the genes for a fatal early onset disorder such as infantile Tay-Sachs (a very rare genetic disorder), then the genes you are born with severely limit your opportunities to live a normal and healthy life. At the other extreme are those persons who inherit the genes associated with exceptional healthy aging – centenarians (age 100 or more) and supercentenarians (age 110 or more). These individuals may enjoy decades more of life free from the diseases that typically afflict humans in their eighties and nineties. Developing new therapies for genetic diseases and/or an intervention that could retard the aging process would constitute enormous medical advances. For this reason alone I believe it is worthwhile to engage with the topics addressed in this book.

The second reason for addressing these issues is that, while our understanding of the role genes play in health, disease and behaviour is still very embryonic and provisional, one often encounters very emotive reactions ("It's eugenics!," "It's unnatural!," "We shouldn't play God!") to the prospect of genetic intervention. And these emotive reactions can hamper the advancement of science and medicine. So I hope this book will help nuance and refine the moral debates on these topics.

And the third and final reason I think it is worthwhile to engage with these topics is the importance of fostering *interdisciplinary* dialogue and debate between the scientists engaged in the kinds of research addressed here and students and scholars in the humanities and social sciences. If the book helps facilitate a better understanding of the relationship between the demands of morality and our knowledge of and ability to manipulate our genes, then I will consider it a success (even if many disagree with the specifics of what I argue).

1

Eugenics: Inherently Immoral?

I What is "eugenics"?

One major reason why critics oppose the prospect of human genetic technologies – whether it be screening technologies, such as pre-implantation genetic diagnosis (or PGD) for parents undergoing IVF, or genetic therapy or genome editing – is that they fear this will take humanity down the same path we visited in the past with unjust eugenics policies. The late nineteenth and first half of the twentieth century witnessed a number of unjust policies (e.g. sterilization of the "unfit") and measures being pursued in the name of "improving the stock" of heredity. This dark episode of our history is what commonly comes to mind when one mentions the word "eugenics."

"Francis Galton, a cousin of Darwin, invented the term [eugenics] and launched a movement to improve the human race, or at least to halt its perceived decline, through selective breeding" (Wikler 1999: 183). Eugenic-like concepts can be found in the ideas of the ancient Greeks. Plato, Aristotle's mentor and teacher, proposed an ideal society that would pursue genetic engineering through selective breeding. The "just" Platonic society was one ruled by philosophers, who possessed knowledge and wisdom and who would govern for the common good. The rest of society, divided into the classes of soldiers and workers, would do the work for which they were best suited. Through censorship (e.g. of poetry) and education of the "guardian class," as well as selective breeding among these class members, the best candidates for the philosopher class could be cultivated and perpetuated. Harmony between the classes could be maintained, argued Plato, if a noble lie was disseminated that some people were made of gold (the philosophers), others of silver and most of bronze. That way people would accept the position in society to which they were best suited.

Most elements of Plato's ideal society will strike us as obviously misguided and authoritarian. Justice, for Plato, meant "doing what you are best suited to do." This ancient construal of the moral virtue of justice was predicated upon the foundational premise that people are *unequal*. Plato's noble lie that people were made from one of the three different elements of gold, silver or bronze clearly shows that his political theory contravened the idea of the moral equality of all persons.

A contemporary interpretation of the virtues of benevolence and justice must be compatible with the basic moral premise that all persons are *moral equals* and, as such, deserve to be treated, in the language of John Rawls (1971), as "separate persons" – persons worthy of respect, equal treatment and the protection of basic rights and opportunities.

How should we define "eugenics"? And what was morally wrong with the eugenic aspirations of Plato and the social movements of the late nineteenth and early twentieth century? Must everything considered "eugenic" necessarily be immoral? These are the questions we consider in this chapter. Virtue ethics is a useful moral framework for helping us understand why the eugenic movements of the past were wrong – namely, they epitomized moral and epistemic *vice*. But, perhaps just as importantly, virtue ethics can help illustrate why eugenics is not inherently immoral. Indeed, justly pursuing eugenic aspirations might actually be required by the virtues of beneficence, justice and the epistemic virtue of adaptability of intellect.

Some have argued that, while "eugenics talk" per se is not wrong, there is something wrong with using its emotive power as a means of circumventing people's critical–rational faculties (Wilkinson 2008). I agree with this point, and the analysis developed in this chapter is one I hope will help reduce emotive reactions to the word "eugenics" and instead bring to the fore the moral and empirical sensibilities needed to determine whether a specific eugenic proposal or aspiration is just or unjust.

Many have defined eugenics in different ways. To help align our moral analysis of eugenics with the social application of virtue ethics developed in this book, I think the most useful way of defining eugenics is one which emphasizes it as a *social movement*. Bertrand Russell (1929: 254) provides the following definition of eugenics, one that defines it as a social movement: "The attempt to improve the biological character of a breed by deliberate methods adopted to that end."

Humans have, since the domestication of animals and agriculture, sought to control and improve the breed of plants and animals they consume and rely upon for labour, sport and leisure. The case of "plant eugenics" is an easy example that satisfies Russell's definition. Humans purposively intervene in the reproduction of plants, for example. Desired phenotypes, such as a beautiful flower or heat-resistant crop, can be achieved through selective breeding or genetic engineering – "a process by which humans introduce or change DNA, RNA, or proteins in an organism to express a new trait or change the expression of an existing trait" (NASEM 2016: 5). The aim of such interventions is to improve the biological character of the plant, to improve its aesthetics, or to aid in feeding the world's growing human population as well as livestock. And a variety of methods are available. These methods are not without controversy. Many critics of genetically modified crops oppose these technologies, either because they believe (contrary to evidence) eating such foods can be unsafe or because they fear they might cause significant harm to the environment.

Russell's definition of eugenics is helpful because it draws attention to two distinct issues, each of which warrants closer consideration: (1) an *end* (i.e. improving the biological character of a breed) and (2) the *means* ("deliberate methods") to achieve (1). Without more details, such as what the specific ends and means are, I believe eugenics is a morally *neutral* aspiration. Some ends are immoral and unjust, whereas other ends are morally laudable. Some means are immoral (even if the ends are morally laudable), and other means are reasonable and defensible. As I have said many times before, the devil really is in the details! I will illustrate this by considering some current public health practices that I believe could be considered "eugenic" but also morally obligatory. And then we will consider the historical eugenic policies that were clearly unjust.

II Folic acid, vaccinations and water fluoridation: eugenics?

Consider, for example, the Centers for Disease Control's recommendation[1] that woman of childbearing age (ages fifteen to forty-five) in the United States should consume 0.4mg of folic acid daily. It is recommended that *all* women in this age group, not only those planning to get pregnant, take folic acid supplements because nearly half of the pregnancies in the United States are unplanned (Finer and Zolna 2016). The goal of having folic acid be taken *en masse* is to prevent two common and serious birth defects – spina bifida and anencephaly. Is this recommendation by a public health agency an example of "eugenics"?

If eugenics is understood as an attempt to "improve the biological character of a breed" by "deliberate methods," then the answer appears to be "yes." It certainly is an interesting case to debate, and it is not obvious the answer is clearly "no" – that prescribing folic acid isn't a case of eugenics. But the important point worth noting is that, even if the recommendation that women take folic acid is "eugenic," it does not mean it is immoral or unjust. In fact, one might take the view, as many public health agencies do, that it is morally obligatory that this recommendation be made. Why? Because it could help prevent serious birth defects. This is beneficial to the prospective parents and their potential offspring. The virtue of benevolence prescribes that a virtuous polity aspire to prevent harm and disadvantage.

Why is the case of recommending folic acid not (at least obviously) morally problematic? I believe there are a few important factors worth emphasizing that are linked to moral and intellectual virtue. Firstly, the CDC simply *recommends* women take folic acid. It is not something that is legally obligatory. Women of childbearing age in the United States are not compelled, under threat of a fine or imprisonment or public shaming, to take folic acid. To go that far would be to contravene the virtue of justice as it would impose unfair burdens on women. Reducing birth defects is important, but it should not be pursued by placing unfair burdens on women of childbearing age.

Secondly, there is a sound *empirical basis* for thinking that taking folic acid is linked to the desired objective – namely, reducing serious birth defects. And as such the measure is an example of *epistemic virtue* rather than vice. The recommended dosage, along with its effectiveness and any potential adverse side-effects, has been extensively studied and the case for prescribing folic acid found to be supported. The CDC does not make the recommendation it does simply on "a hunch" or because it wants to increase the monetary revenue of the pharmaceutical companies producing folic acid.

Thirdly, the desired goal in this case is a morally laudable one – reducing the prevalence of serious birth defects. Such an aspiration is compatible with *the virtue of benevolence*, which instructs us to prevent harm and disadvantage when possible and reasonable to do so. So invoking moral and epistemic virtues can help us understand why the CDC's recommendation is not objectionable. Indeed, a strong case can be made for arguing that it is morally obligatory. It simply encourages women of childbearing age to undertake the minor burden of the daily consumption of a pharmaceutical that could promote the health of her offspring, whether it was a planned or unplanned pregnancy. But what about the unjust eugenic policies of the past? They exemplified, I shall argue shortly, both moral and epistemic *vice* (not virtue).

Before turning to past eugenic policies, let us consider a second contemporary public health example – vaccinations. Children in developed countries are protected from many infectious diseases that kill children in less developed countries because the government can afford, and actively pursues, nationwide initiatives of vaccines to reduce the odds of an outbreak of vaccine-preventable diseases such as pertussis, mumps and measles. Such an aspiration certainly seems to fit Russell's first condition for something to count as "eugenics" – it seeks to improve our biological character by making us less susceptible to infectious disease.

And deliberate methods are adopted to try to realize this aim. For example, in the United States there are state laws that require vaccinations for schoolchildren and children in day-care facilities. So healthy children are required to receive a medical intervention that is neither 100 percent effective nor 100 percent safe.[2] Vaccinations have some risks of harm, and side-effects may include mild problems such as a low-grade fever or more serious complications (these are very rare). However, the overall risks are very small compared with the risks all children would face without vaccinations. And this is why vaccinations are such an important part of public health initiatives.

The benefits of "herd immunity" – achieved when a sizeable portion of the population is immunized, protecting most of its members from the disease – far outweigh the costs and risks associated with vaccinations.

Are state-sponsored vaccination initiatives such as those pursued in the United States "eugenics"? Again, I think one could persuasively argue, "yeah, they are!" Critics often make precisely this point to try to weaken support for vaccination programs. But just because it is eugenics does not mean it is morally objectionable. Such programs seek to improve the biological character of children by enhancing their immune system so they are less vulnerable to infectious disease. We do not take the view that the biology humans have inherited from the process of evolution by natural selection is "good enough," and that we should just tolerate higher rates of infant and child mortality rather than tamper with our "natural" biology. Instead, we pursue the goal of improving the health of the population by mandating that healthy individuals be subjected to medical interventions that may not be 100 percent safe or 100 percent effective. But, just because vaccination programs are "eugenic," that does not mean they are immoral or unjust. In fact, the provision and (reasonable) enforcement of vaccinations can be considered *morally obligatory* (not simply morally permissible). Labelling something "eugenics" does not tell us much about the moral character of the practice in question. More details are required – namely, whether the eugenic policy is one that violates moral and epistemic virtue.

One last contemporary public health measure which seeks to "improve the biological character of a breed by deliberate methods adopted to that end" is water fluoridation. Since the 1950s, when evidence concerning its benefits became clear, water fluoridation has been widely used in the United States and many other developed countries. The CDC estimates that, by keeping low levels of fluoride in the mouth all day, fluoridated water reduces tooth decay by 25 percent (beyond what the use of toothpaste and mouth rinses can achieve) among children and adults.[3] But, one might ask, is water fluoridation "eugenic"? Consider the finding noted by the CDC: "In communities with water fluoridation, school children have, on average, about 2 fewer decayed teeth compared to children who don't live in fluoridated communities."[4] I do not think it is a far stretch to describe water fluoridation as an intervention that aspires to "improve our biology" and one that is sought by "deliberate methods." It is thus a "tooth enhancement," helping to insulate people from the problems of decay to which we are susceptible with the "normal tooth functioning" provided by evolution by natural selection.

With water fluoridation, the measures invoked to achieve the aim of improving dental health are non-coercive. No one is forced to consume fluoridated water (or to brush their teeth). One can "opt out" of water fluoridation by choosing to drink only bottled water that is not fluoridated. There is of course an additional cost associated with this option. But the reason water fluoridation is not a morally objectionable eugenic policy stems from the fact that it is a public health measure consistent with moral and epistemic virtue rather than vice. It aims to benefit people (by preventing tooth decay), and the means employed to achieve that goal are rationally connected to that goal and do not impose unfair burdens on persons (e.g. violating basic rights and freedoms).

The examples of folic acid, immunizations and water fluoridation are, admittedly, debatable cases to describe as "eugenics." The critic might object to my making such comparisons by noting that new germline genetic modifications – interventions that alter the genes of sperm and egg (and thus impact future generations) – *permanently* alters the biology of the human species. Such an intervention might apply genome editing to embryos for implantation with IVF or a germline gene therapy to a person suffering a heritable disease. And because such an intervention is permanent, they might contend, it warrants extra special ethical concern. My response to this line of objection is to note that, should it be possible for us to utilize genome-editing tools to alter the germline, then it is also likely that we could reverse such edits if they resulted in negative consequences we did not foresee. Purposively altering, even permanently, the biology of our species is not itself necessarily a worry. I do not see why the prospect of eliminating serious genetic disorders should concern us. This is why we undertake such medical research in the first place – to eliminate such diseases! Permanently. Altering our germline would be a worry only if this alteration brought about some (unforeseen) negative consequence. No doubt some humility for recognizing the limitations of our ability to foresee such adverse outcomes is certainly warranted. But I do not see why we should presume that a germline genetic intervention would need to be permanent and irreversible. If we can successfully intervene to alter the genome we inherited from evolution by natural selection, presumably we could also edit the genome we purposively modified.

The reason I invoked the examples of folic acid, immunizations and water fluoridation was not to wrangle with semantics about the word "eugenics." But I believe Russell's definition is instructive because it compels us to reflect on the variety of ways we already try to improve the biology of our species. If Martians arrived on Earth and asked why we prescribed folic acid to women of childbearing age, we could plausibly answer "To improve the biology of newborns, so they have a lower risk of serious birth defects." If asked why we provide vaccinations to children, we could answer "To improve our biology so people are less at risk of infectious disease." And if asked why we put fluoride in the water, we might reply "To improve our biology by making our teeth less susceptible to decay." In all three cases we have decided to utilize knowledge to improve upon the biology given us by evolution through natural selection. While there are critics of all three public health measures, these measures are sound because (1) they seek to prevent something harmful (e.g. serious birth defects, infectious disease and tooth decay), as prescribed by the virtue of *benevolence*, and (2) they aspire to do so by measures that are rationally connected to their objective (as required by the *epistemic virtues*) *and* (3) they are consistent with the requirement that we treat persons as free and equal (as required by the virtue of *justice*). These measures do not propose sterilizing anyone or throwing people in jail for their procreative decisions. Such proposals would clearly violate the individual rights protected by the virtue of justice (3).

III The (unjust) eugenics of the past

The eugenic aspirations of the late nineteenth and early twentieth century clearly contravene the demands of moral and epistemic virtue. This becomes most evident when we consider that many of these policies took a prejudiced interpretation of the *end* to pursue with "improving our biological character." Racial hygiene, for example, is an aspiration of moral *vice*, not virtue. It is predicated upon the assumption that some people are naturally superior to others, and this contravenes the virtue of justice and its prescription that all should be treated fairly and impartially. Forceful sterilization of the "unfit" is another example of moral vice, imposing unjust burdens on individual persons in the name of "improving the stock."

Furthermore, measures that sought to curb criminal behaviour, for example, by discouraging breeding among those deemed "undesirable" (such as the Jukes)[5] were not based on sound empirical, scientific findings. As such, their *means* to achieve the end of improving our biology were not rationally connected to this objective. And as such they exemplify epistemic *vice* rather than virtue. Intellectual virtue requires that we have the ability to recognize the salient facts and have both a sensitivity to details and intellectual humility. These were all absent from the unjust eugenic aspirations of the late nineteenth and early twentieth century.

And, finally, invasive measures such as discouraging or prohibiting some from reproducing violates the requirements of (3) the virtue of justice. We know, for example, that the age of both the mother and the father can increase the health risks for their potential children. A polity that sought to "improve our biological character" by prohibiting anyone over age forty from having children, or even sexual relationships, would clearly be unjust. It would impose a significant burden (e.g. violation of reproductive freedom) on individual persons that cannot be justified by an appeal to the collective benefits of the polity as a whole.

These lessons from the past ought to inform our moral compass today in terms of what we see as permissible, and impermissible, societal aspirations to improve our biology. If carriers of germline genetic disorders were compelled to reproduce only through IVF with gene-edited embryos, then we would be guilty of exercising moral *vice* in a fashion similar to compelling the sterilization of the "unfit." This is so because we would fail to treat such parents with the respect and deference that ought to be given to free and equal persons when it comes to their reproductive freedom. People should have the freedom to decide with whom they wish to procreate, and how (e.g. naturally or through IVF).

IV Education and exercise: eugenics?

I would like to suggest, perhaps provocatively, that Russell's definition of eugenics also applies to two other common measures we
typically think of as morally obligatory – education and exercise.
If we utilized brain-imaging technology to compare the brain of an
average fourteen-year-old from 50,000 years ago to the brain of
an average fourteen-year-old in the United States today, we would
observe disparities in their physical structure. Regions such as the
cerebral cortex, for example, would have noticeable differences. And
the nerve cells would also be distinct in terms of how efficient they
were. The stimuli of the external world helps shape the development
of the brain, our most complex of organs.

Learning makes nerve cells more efficient or powerful, and contemporary learning environments are complex environments that
help "improve the biological character of the breed." The learning
environment typical of early hunter-gatherer societies was much less
complex than those typical of today's world. The systems of primary
and secondary education that children receive impact their brain
development in fundamental ways. Children today are exposed to
many unique stimuli that alter their brain development – such as
creativity, socialization, learning new information, memorization, etc.
Public schooling and education is a policy initiative of all developed
countries that hope to flourish economically as well as provide the
stability needed for democratic traditions and practices and provide
equality of opportunity for all.

Granted we do not typically justify the provision of public education in eugenic terms such as "improving our biological character."
One might defend such measures on the grounds that children
have a right to learn and develop, or that public education can
help promote equality by mitigating the unequal effects of the
social lottery of life. Or economists might emphasize the importance of public education in helping a polity retain a competitive
edge in a globalized world, etc. Nonetheless, we could redescribe
all of these same rationales in the language of "improving our
biological character": that children have a right to a healthy,
creative and developed brain, that all children should have equal
opportunity to realize their cognitive potential, or that a polity

that excels in cognitive development will succeed in a competitive world.

Once we appreciate the point that learning alters our biology (for the better), we realize that it is not inconceivable to see that even diet and exercise are "eugenic" in that they aspire to improve our biological character. The food we consume and our lifestyle (sedentary or physically active) actually affect how genes are expressed. Insights from epigenetics, which we will examine in detail in chapter 4, have revealed this important and complex "environment–gene" interaction. People visit the gym regularly to try to improve their biology – to build stronger muscles, reduce body fat, strengthen their bones, etc. There are countless "fad diets" tried every year to help people lose weight. Our metabolism changes as we age, but exercise and sensible eating habits can help improve our health prospects across the lifespan.

What the examples of diet and exercise show is that even things such as government-proscribed "healthy diets" and lifestyles can be considered "eugenic" in the way that Russell defined that term. They seek to improve the biological character of a population through *deliberate* methods (e.g. cycle lanes in busy cities, protection of parks for walking in nature, etc.). But just because such measures might (arguably) be called "eugenic" does not mean they are morally objectionable. When eugenic measures exemplify moral and epistemic *virtue*, rather than vice, they are morally obligatory rather than simply morally permissible.

Prescribing healthy diets and exercise can benefit people by reducing their risks of disease. Such prescriptions are not based on simple "hunches" or "intuitions" but on sound empirical findings backed by scientific studies. And the *means* employed to achieve the aim of better health (such as education about how the food we eat impacts our health) is not coercive or morally objectionable. The end of better health is a just aspiration, and employing the means of education and encouragement to achieve that aim is fair and reasonable. A polity that encourages healthy diet and exercise for all its citizens comes closer to realizing the moral and epistemic virtues than a polity that remains indifferent to the suffering arising from an epidemic of obesity and the mental health problems associated with a sedentary lifestyle.

Prescribing daily physical exercise of course faces a serious limitation – lack of compliance. Everyone knows regular exercise is good for their physical and mental health, and yet millions of people do not comply with the recommended requirements for physical activity. Why? The scarcity (or perceived scarcity) of time seems to be one obvious factor. People are very busy. Parents have children to feed, kids' activities to attend, food to cook and the house to clean, not to mention the pressing demands of work – work that, for many, includes lengthy hours sitting at a computer, for example. There might also be economic considerations at play – a parent who can afford a gym membership with day-care for toddlers will have an advantage over someone who cannot afford the gym membership or is a single parent with young children to look after.

What if there were a way to gain even greater health benefits than might be attained by regular exercise and sensible eating in a way that would likely achieve even higher compliance rates? This might sound like science fiction today, but the field of biogerontology might make such an intervention a reality. This will be explored at great length in chapter 6.

V From unintentional genetic modification to intentional genetic modification

Why don't we generally take the view that the social movement of offering public education, for example, is a "eugenic policy"? I think there are at least two reasons: the first concerns its purported aim and the second the means it invokes to realize that aim.

Firstly, schooling is a form of what we might call "environmental engineering." Environmental engineering involves using knowledge and technology to improve the environments in which we live, most (if not all) of which alter human biology in some form or other. And this is the case because our biology is shaped and influenced by both nature (e.g. the genes we inherit) *and* nurture (e.g. education, the diet we consume, the relationships we experience, etc.). The world is a hostile place for all species, including humans. Historically humans did not survive long given the threats of infectious disease, poverty, violence, etc. Various forms of environmental engineering have permitted humans to improve their life prospects, many of which alter our biology (for better or worse).

The agricultural and industrial revolutions, for example, improved food production and the materials needed to build homes and clothing. Automobiles increased our ability to travel long distances quickly. The printing press, the telephone, computers and the internet have increased the dissemination of knowledge and connectivity of humans (with social media such as Facebook and Twitter). Some of these have obvious impacts on our biology. The availability of cheap, high calorie foods, coupled with the inactive lifestyles of a culture heavily reliant on automobiles for transportation and long working hours sitting at a desk, means that obesity has become a health problem on a scale that would not have been possible in hunter-gatherer societies. Indeed, some predict that life expectancy in the United States might actually decrease in the future as a result of the rise of childhood obesity (Olshansky et al. 2005).

So, one reason that education might not typically be classified as "eugenic" stems from our failure to pay attention to the fact that environmental engineering alters our biology. But it does. A second reason we don't typically think of education as "eugenic" concerns its *means* – when "eugenics" is raised as a term of disapprobation we think of state coercion, for example sterilizing persons or encouraging "the best" to bear more offspring. Education is not about state coercion; it does not involve violations of reproductive freedom. But it is worth bearing in mind that public education, at least during childhood and early adolescence, is *compulsory* (subject to some exceptions, e.g. home schooling) in developed liberal democracies. Even so, we see the provision of public education as not only morally permissible but morally obligatory. And a virtue ethics lens helps us make sense of why we have such strong convictions about this, because education is an integral part of moral and epistemic virtue. Education helps enhance the "menu of options" for the citizenry (helping them develop as autonomous persons, etc.), it improves the economic prosperity of a polity by ensuring citizens are skilled and literate, and it facilitates the exercise of the intellectual virtues (e.g. social communication, attention to details, etc.) and moral virtues (e.g. empathy and understanding for others).

The purpose of addressing the term "eugenics" and the examples of folic acid, vaccinations, water fluoridation, education, exercise and diet was twofold. Firstly, describing a measure as "eugenic" because it aspires to improve our biology by deliberate means does not mean it is a morally invalid aspiration. The reason why I focused on public health measures that intentionally aim to improve human biology – vaccinations, education and exercise – is to open our eyes to the reality that we already seek to improve upon the biology given us through evolution by natural selection. And there are significant costs, even risks, associated with these measures. But we don't consider them immoral. In fact, we see them as morally obligatory, so much so that their provision is a requirement of both benevolence and justice.

Secondly, I address the examples of folic acid, vaccinations, water fluoridation, education, exercise and diet to illustrate the important point that what really matters in each of these cases are the *aims* of such policies (e.g. to prevent disease and disability, promote equality, promote health, etc.) and the means by which those *aims* are pursued. We see these expensive and potentially risky interventions as important policies to strive for in order to ensure individuals and polities flourish. And the *means* invoked to pursue these aims can vary, from the recommendation to take folic acid to legislated vaccination programs, mandatory education for all children, and disseminating knowledge about healthy eating and exercise.

And yet, when it comes to the prospect of intentionally altering the biology of humans via gene therapy or genome editing, all of a sudden people's attitudes can change, and they come up with emotive responses such as "That's eugenics!," "Don't play God!," or "That's unnatural." I hope this chapter has effectively illustrated the folly of such emotive replies.

In "Breaking evolution's chains: the prospect of deliberate genetic modification in humans," Russell Powell and Allen Buchanan (2011) implement a very helpful distinction between *unintentional genetic modification* (UGM) and *intentional genetic modification* (IGM) that can help us further understand the folly of equating "eugenics" with "injustice."

Evolution by natural selection, which will be explored in greater detail in the next chapter, is an example of unintentional genetic modification (UGM). Genes that are beneficial, that have an adaptive benefit in terms of increasing health and reproductive success in particular environments, are passed on with greater frequency than genes that are neutral or maladaptive. If the latter cause mortality before the age of reproduction, they are not passed on to future generations. In *The Long Tomorrow*, the evolutionary biologist Michael Rose explains how the UGM of natural selection impacts our health prospects across the lifespan:

> Natural selection discards bad genes, genes like those that cause fatal childhood progeria. Bad genes cause these effects by producing inborn errors of metabolism: letting toxins accumulate, impairing brain function, and so on. Many of the diseases that kill infants are the products of such bad genes ... Natural selection keeps genes with such devastating early effects rare, because the afflicted individuals die before reproducing. Bad genes destroy themselves when they kill the young ... But at later ages, the force of natural selection becomes weak. It leaves genes with late bad effects alone, because natural selection has stopped working. Its force has fallen toward zero. Bad genes that only have late effects will not be removed by natural selection. They can accumulate. There is no more automatic Darwinian screening. (Rose 2005: 42)

Darwinian screening is a form of unintentional genetic modification (UGM). And it is far from perfect. It doesn't care about the virtues of benevolence or justice or human wellbeing because it cannot care about anything. Culture is also a form of UGM. Patriarchy, slavery, war, etc., influence who reproduces and who survives, which all affect the biology of a population. Public health measures such as sanitation and immunizations also unintentionally modify our evolution as a species, as they result in keeping some people alive long enough to become parents, whereas, if we remained in the environments, and with the knowledge typical of, early human civilizations, many of them would have died before reaching the age of sexual maturity. Is UGM inherently good? "No." Is UGM inherently bad? "No."

Now if we shift to the prospect of intentional genetic modification (IGM), and we ask if IGM is inherently good or bad, unfortunately many will rush to answer that it is morally bad. Why this response? Because the only examples of IGM most people will consider are past eugenic policies such as sterilization of the "unfit," which were indeed unjust. But I do not think IGM is inherently unjust. If we can develop safe and effective ways to prevent disease and promote health and happiness, then IGM can be considered not only morally permissible but morally obligatory, because it is required by the virtues of benevolence and justice (just as vaccinations and water fluoridation are).

The prospect of genetic engineering offers us the opportunity to shape human biology and human evolution in ways that go beyond what evolution by natural selection and environmental engineering have permitted. If new genetic technologies such as gene therapy, genome editing or an aging intervention prove to be safe and effective ways to promote more health, equality and happiness, it would be folly to forfeit such innovations because of some romanticized commitment to remaining committed to UGM as the only morally legitimate driving force to shape human biology and evolution.

Discussion questions

1 If we put aside the issue of "genetic engineering" for the moment, do you think the aspiration to "improve the biological character" of humanity via other types of interventions (e.g. public health measures, education, etc.) is morally laudable (even if potentially problematic)?

2 Can the virtue ethics framework (employing both moral and epistemic virtue/vice) help us diagnose the ills of past eugenic practices and prescribe how we can harness the potential benefits of genetic intervention without repeating the atrocities of the past?

3 Are you in favour of a potential shift from "unintentional genetic modification" towards "intentional genetic modification"? What fears, concerns, risks, benefits, aspirations, etc., come to mind when you consider such a prospect?

2

The Genetic Revolution: A Snapshot

I Introduction

We are living in the midst of what I refer to elsewhere (Farrelly 2016) as the "genetic revolution." Genes are the fundamental physical and functional units of heredity; they "specify the proteins that form the units of which homoeostatic devices are composed" (Childs 1999: 5). The genetic revolution constitutes significant progression (1) in our understanding of our biology and the role genes play in the development of health, disease, intelligence, behaviour, etc.; and (2) in novel innovations and interventions, from diagnostic tools (e.g. genetic testing and pre-implantation genetic diagnosis (PGD)) to therapeutic interventions (e.g. gene therapy or genome editing to prevent or treat specific diseases such as single-gene disorders) and possibly "enhancing" interventions (e.g. a drug that retards aging or a memory modification drug that boosts our "psychological immune system") that could be realized by the knowledge acquired in (1).

II Historical thinkers

The theoretical foundations of the genetic revolution really took root in the nineteenth century. No individual scientists are alone responsible for the revolution, but two important figures that were significant catalysts for the revolution are worth highlighting – Charles Darwin

(1809–1882) and Gregor Mendel (1822–1884). These two think-ers exemplify the "intellectual virtues," and their contributions to science and humanity are important for us to emphasize, as such virtues must inform the moral virtues of benevolence and justice if we are to realize *phronesis* (practical wisdom) in the twenty-first century.

Darwin was a naturalist who, in 1835, travelled around the world by ship (HMS *Beagle*) on a five-year journey. Making meticulous notes of the different species he encountered on his way, Darwin published his reflections in *On the Origin of Species* in 1859. Species (gradually) evolved, argued Darwin, and the driving force behind this evolution was natural selection. This secular view of life was a revolutionary break from the established theological perspectives of life on the planet.

Darwin exemplified the "intellectual virtues," especially an "adap-tive intellect" and the "detective's virtues." The latter prescribes that we think of coherent explanations of the facts. Darwin observed the diversity of species he encountered, together with the diversity of traits and characteristics (e.g. the variation in the shape and size of finches' beaks) within a species, and formulated the only coherent explanation that explained why such diversity existed – evolution by natural selection.

Gregor Mendel's meticulous study of genetic inheritance in the pea pod exemplified the intellectual virtue of having a sensitivity to details. Mendel cross-bred pea plants over many generations, and from his observations he conjectured that there must be heredity determinants (genes), and that genes are in pairs.

III Contemporary science

A further major breakthrough in our understanding of genetics came in the middle of the twentieth century when, in 1953, James Watson and Francis Crick discovered the double helix structure of DNA, thus helping give rise to modern molecular biology. The first officially sanctioned human gene therapy took place on September 14, 1990. Ashanti DeSilva was among the first children to receive somatic-cell gene therapy; she "was the first of two children to receive a dose of her own cells in which a functioning counterpart of her malfunctioning gene had been previously inserted" (Walters and Palmer 1997: 17).

The subjects of the earliest gene therapy experiments were children who suffered from a rare genetic disease called adenosine deaminase (ADA) deficiency.

Tragically the first person to die from a gene therapy experiment was eighteen-year-old Jesse Gelsinger: "Gelsinger had ornithine transcarbamylase (OTC) deficiency, a metabolic disorder that affects 1 in 40 000 newborns by impeding the elimination of ammonia. Most of these babies become comatose within 72 hours of birth and experience severe brain damage. Half die within a month of birth, and half of the survivors die by age 5" (Sibbald 2001: 1612). Gelsinger participated in an experimental gene therapy trial at the University of Pennsylvania. Within days of undergoing the procedure his organs began to fail, and he died shortly after receiving the therapy. His death is believed to have been triggered by a severe immune response to the adenovirus carrier. Jesse's was the first reported death that was directly attributable to a gene therapy experiment, and it sparked a backlash. Bill Frist, a United States Republican senator, called for hearings on the oversight of the use of gene therapy, and these hearings began in February 2000.

The late 1990s were a time of much excitement and anticipation, as two rival teams were racing to sequence the human genome. The "race" ended in February 2001, when the two teams produced their draft versions. The publicly funded Human Genome Project published its results in the scientific journal *Nature* (International Human Genome Sequencing Consortium 2001), while the private US firm Celera Genomics published its results in *Science* (Venter et al. 2001). "The estimated cost for generating that initial 'draft' human genome sequence is ~$300 million worldwide, of which NIH provided roughly 50–60%."[1] Craig Ventor, one of the founders of Celera Genomics, was the first person to have his personal genome sequenced, in 2007 (Levy et al. 2007), at an estimated cost of $150 million (Watman 2008). The cost of sequencing a genome has been reduced dramatically since then. James Watson had his genome sequenced in 2008, with rapid-sequencing machines, at a cost of only $1.5 million. The much anticipated goal of a $1,000 genome is now well within reach as the cost, as of 2015, was estimated to be below $1,500,[2] though there is controversy surrounding the quality of lower-cost sequencing techniques.

What is the potential value of being able to sequence a person's genome for only $1,000? It might permit us to adopt a more personalized approach to medicine. At the moment, when a patient is diagnosed with a medical problem (e.g. hypertension, high cholesterol), they are prescribed the standard medication that typically helps manage the condition. However, for some patients the standard medication might not be effective. Indeed, it might even put them at risk of serious side-effects. Personalizing medical treatment (and preventative measures) based on a person's genome might help us promote health in a more effective and safe manner. The realization of a $1,000 genome would be a monumental achievement if this were the case.

The number of genetic tests available dramatically increased as the twentieth century came to a close, from the handful developed decades earlier, for PKU disease, cystic fibrosis (CF) and Duchenne muscular dystrophy, to tests for thousands of rare and more common genetic diseases. The potential benefits can vary depending on the type of testing and condition. If both prospective parents know they are carriers of the CF gene, this means there is a 25 percent chance that a child would develop CF. That information might lead the parents to decide to pursue IVF with a non-carrier donor, utilize PGD to implant an embryo that does not have the CF gene, adopt or not have children, or make plans in the event of having a child with special healthcare needs. A child born with PKU disease can be prescribed a special diet to avoid the harms of the condition. Whereas there is no cure for Huntington's disease, those who choose to undergo testing for HD might wish to plan their future based on such information.

"Preimplantation genetic diagnosis (PGD) permits genetic testing before the transfer of embryos to the mother, which has distinct advantages in establishing pregnancies unaffected by tested disorders and without the potential need for pregnancy termination" (Kuliev and Verlinsky 2005). The procedure involves first obtaining eggs and fertilizing them in vitro. After a few days of developing in the laboratory, cells can be removed from each embryo and be tested for genetic or chromosomal abnormalities, as well as sex. With the information yielded by the tests, unaffected embryos, or the embryo of the desired sex, can be implanted into a woman's uterus. When successful, the procedure results in a pregnancy and the birth of a child not affected by the abnormality or of the desired sex. Because many genetic disorders are sex-linked – for example, Duchenne muscular dystrophy primarily affects boys – in this instance selecting for a girl would be a way to improve the odds of implanting an unaffected embryo.

Does removing a cell from the embryo pose any health threats to the children born from IVF who undergo PGD? The answer appears to be "no" (Liebaers et al. 2010), though it is a topic that still needs to be monitored closely.[3] In addition to testing embryos for early onset disorders, it is possible to screen both for late onset conditions and sex for non-medical purposes (e.g. family balancing). The ethical issues (e.g. scope and limits of reproductive freedom) surrounding the use of PGD for medical and non-medical purposes will be examined in chapter 5.

IV Behavioural genetics

The first full genome scan of sexual orientation in males was published in the March 2005 issue of *Human Genetics* (Mustanski et al. 2005). And, while sexual orientation is a complex phenotype (there is no "gay gene"), that study identified several chromosomal regions and candidate genes for further exploration. Over the past number of decades the field of behavioural genetics has also taken off, with fascinating discoveries concerning the role that heredity plays in criminal behaviour (Tiihonen et al. 2015), voting (Fowler and Dawes 2008) and parental care (Pérusse et al. 1994). Behavioural genetics offers such a diverse array of insights and unique ethical considerations that it would warrant a separate book all on its own. So we, or rather I should say *I*, cannot do justice to these important and timely issues here. Instead I shall just briefly mention the issue of criminal behaviour, parental investment and intelligence. And in chapter 7 we shall consider happiness, memory and moral behaviour.

The prevention, and treatment, of criminal wrongdoing is a major concern for a virtuous polity. In liberal democracies such as America criminal punishment serves many distinct purposes – we threaten potential criminal offenders with fines and prison to *deter wrongdoing*; we incarcerate dangerous offenders to *protect the public*; we punish (and threaten punishment) to *educate*, not just offenders, but society as a whole (e.g. concerning which actions are morally right and wrong); and we punish to satisfy the demands of *retributivism*.

This final reason prescribes that we punish because wrongdoers *deserve* to suffer. Michael Moore (1987), a prominent proponent of retributivism, claims that the moral culpability of an offender gives society a *duty* to punish. According to such a view, a just society thus has an obligation to set up institutions so that retribution is achieved. And John Rawls claims "That a criminal should be punished follows from his guilt, and the severity of the appropriate punishment depends on the depravity of his act" (1999: 22). The fulfillment of the demands of desert is, according to retributivism, an *intrinsic* good. Jean Hampton claims that retributivists see punishment as performing the rather metaphysical task of "negating the wrong" and "reasserting the right" (1984: 215).

In contrast to retributivism, the moral education theory takes the view that the punishment of wrongdoers aims (primarily though not exclusively) to benefit the offender. The rationale for punishment is thus *paternalistic*: punishment communicates a moral message to the offender, a message they need to hear in order to repair moral defects in their personality. This account has been championed by a variety of different thinkers. Protagoras makes a speech in Plato's dialogue *Protagoras*[4] which succinctly captures this account of punishment:

> In punishing wrongdoers, no one concentrates on the fact that a man has done wrong in the past, or punishes him on that account, unless taking blind vengeance like a beast. No, punishment is not inflicted by a rational man for the sake of the crime that has been committed (after all one cannot undo what is past), but for the sake of the future, to prevent either the same man or, by the spectacle of his punishment, someone else from doing wrong again. But to hold such a view amounts to holding that virtue can be instilled by education; at all events the punishment is inflicted as a deterrent. (*Protagoras* 324a; Guthrie 1956)

In their study on the genetic background of extreme violent behaviour in Finnish prisoners, Tiihonen and his colleagues (2015) found that a monoamine oxidase A (MAOA) low-activity genotype (contributing to low dopamine turnover rate) and the CDH13 gene (coding for neuronal membrane adhesion protein) were associated with extremely violent behaviour. Furthermore, the researchers estimate that at least about 5 to 10 percent of all severe violent crime in Finland is attributable to these genotypes. Such findings raise fascinating questions concerning how a virtuous polity ought to respond to criminal wrongdoing. Do findings of this kind compel us to reconsider the role of desert and education in punishing some offenders? Are wrongdoers who possess the genes associated with violent behaviour any less morally culpable because of their biology? Is long-term incarceration an effective way of communicating the wrongs of such behaviour to such criminals? Can insights from genetics help us rehabilitate such offenders? Understanding the role genes play in criminal behaviour could have important implications for criminal justice.

The second aspect of behavioural genetics I will refer to briefly concerns parental investment. "Studies across a wide variety of taxa have established that variation and parental care is heritable and partly mediated by genetic mechanisms" (Champagne and Curley 2012: 304). From nesting behaviour to the rates of feeding offspring in non-humans, to the study of twin adults and their investment in parenting, it is clear that genes certainly play an important role in such care. This has potentially significant implications for the demands of benevolence and justice. Parents profoundly influence their children, from their diet to their behaviour, personality and interests. In addition to the education and income of one's parents, the genes they possess influence their parental care, which in turn influences the life prospects of their offspring. Children do not choose their parents, and thus some will be raised in families that are unequal not only in socio-economic terms (e.g. less parental wealth and educational attainment) but also in terms of the natural inclination to invest more in care. This inequality of opportunities raises questions about the demands of justice in the future, as we learn more about the specific genes that influence parental investment. If it should become possible to modulate particular genes to increase parental investment in those who lack the genetic predisposition, does justice require we encourage interventions that could mitigate this natural inequality for the sake of potential offspring? Would such a moral duty simply be an extension of the duty to mitigate other types of inequality (socio-economic) that influence the opportunities children have that are shaped by family life? I do not think there are easy or obvious answers to these questions. The virtuous polity may have to grapple with such questions in the years to come.

The first genome-wide association study (GWAS) on childhood intelligence (age range six to eighteen years) was published just a few years ago (Benyamin et al. 2014). Intelligence is a complex phenotype, and providing a comprehensive definition of "intelligence" is contentious. Here is an influential characterization:

> Intelligence is a very general mental capability that, among other things, involves the ability to reason, plan, solve problems, think abstractly, comprehend complex ideas, learn quickly and learn from experience. It is not merely book learning, a narrow academic skill, or test-taking smarts. Rather, it reflects a broader and deeper capability for comprehending our surroundings – "catching on," "making sense" of things, or "figuring out" what to do. (Gottfredson 1997: 13)

The importance of intelligence in terms of an individual's, and a polity's, opportunity to flourish could not be overstated. And "high intelligence is precious human capital for advancing and maintaining society in the information age" (Shakeshaft et al. 2015: 123). "Catching on" and "making sense of things" can help a person better succeed in romantic relationships, work and parenting. A society's ability to "learn from experience," "reason" and "solve problems" dramatically influences its economic prosperity, political institutions, medical breakthroughs and technological innovations. Some estimate that the heritability of intelligence increases from about 20 percent in infancy to perhaps 80 percent in later adulthood (Plomin and Deary 2015).

From complex areas of inquiry, such as neuroscience and cancer research, to quantum mechanics and public policy, the prospect of possibly being able to improve the intelligence of humans via genetic engineering would mean humanity could potentially make improvements in a multitude of domains where our limited cognitive functioning constrains moral progress.

V Genetic discrimination, gene patents and gene therapy

The rapid increase in genetic testing over the past few decades has brought with it the concern that new information could be used to disadvantage people who are at higher risk of genetic disorders. Concerns about the prospect of information of this kind being used to deny people employment or health insurance led to legislation to guard against such risks. In 2008 the United States passed the Genetic Information Nondiscrimination Act, and many other countries have followed suit. In fact, as I write these words, the Canadian government is set to bring Bill S-201 Genetic Non-Discrimination Act into law, though Prime Minister Justin Trudeau has requested the Supreme Court provide their advice on the constitutionality of the private member's bill.

By the end of 2000, two years before the final publication of the human genome in 2003, more than 25,000 DNA-based patents had already been issued (Cook-Deegan and McCormack 2001: 217). These patents grant the rights to *exclude others* from making, using, selling, offering for sale or importing patented items for twenty years. Are such rights *morally* justified? Critics argue that patents can impede the development of new medical interventions. Michael Heller and Rebecca Eisenberg, for example, published a paper in *Science* entitled "Can patents deter innovation? The anticommons in biomedical research," in which they argued that "a proliferation of intellectual property rights upstream may be stifling life-saving innovations further downstream in the course of research and product development" (Heller and Eisenberg 1998: 698). This stems from the problem they call the "tragedy of the anticommons." This problem occurs when multiple owners each have a right to exclude others from a scarce resource and yet no one has effective control over the resource. When this happens, argue Heller and Eisenberg, "collecting rights into usable private property is often brutal and slow" (ibid.). In 2013 the Supreme Court of the United States ruled, in the case of the *Association for Molecular Pathology* v. *Myriad Genetics*, that "naturally occurring" human genes cannot be patented in the United States because DNA is a "product of nature."

Despite major setbacks like the tragic death of Jesse Gelsinger, gene therapy clinical trials have forged ahead for a variety of different diseases. Worldwide there are approximately 2,500 clinical trials,[5] the majority (nearly 64 percent) taking place in the United States. Nearly 65 percent of these experimental trials are for cancer and only 7.5 percent for infectious diseases. To ensure such interventions are empirically and morally sound, medical research is rigorously regulated through different "phases." Over half of the world's gene therapy trials are phase I clinical trials. It is important to point out that most (estimates suggest around 90 percent; Smietana et al. 2016) potential drugs that enter phase I drug experiments eventually fail to succeed. That is, they fail to prove to be a safe and effective intervention that does a better job than existing products and interventions. And to develop and gain marketing approval for a new drug in the United States was recently estimated to be $2.558 billion (DiMasi et al. 2016). Clinical trials are risky and extremely expensive endeavours, but they are also essential to the progress of science and medical research.

Renewed enthusiasm about the prospects of our being able to intervene directly in our genes has grown in just the past few years, as more efficient and reliable ways to make precise, targeted changes to the genome of living cells have proved possible with CRISPR, which permits medical researchers to edit parts of the genome by removing, adding or altering sections of DNA sequence. In 2016 a Chinese group became the first to inject a person with cells that contained genes edited using the revolutionary CRISPR-Cas9 technique (Cyranoski 2016). The patient suffered an aggressive lung cancer. And as of writing, in early 2017 the United States is set to start their first clinical trial for gene editing as well. So the race to be the first to develop gene-edited cells that can be used in clinics is now in full force.

VI Clinical trials

How does a society ethically permit experimenting on humans, whether it be gene therapy or gene editing or any other form of medical intervention? The FDA website provides a useful summary of the different phases of clinical trials.[6] I will provide a brief summary of that summary here, as the FDA demonstrates the complex issues that must be addressed when trying to exercise the "intellectual virtues" in the context of undertaking medical research that could help us better realize the demands of benevolence and justice.

Before any analysis is conducted on humans, research takes place in the laboratory setting on animals. In pre-phase trials, tests are conducted on species such as mice. Mice reproduce quickly and can be maintained relatively cheaply. Are experiments on mice good models for human medical research? "On average, the protein-coding regions of the mouse and human genomes are 85 percent identical; some genes are 99 percent identical while others are only 60 percent identical."[7] In the United States, medical research is regulated by federal laws that seek to ensure the animals are treated humanely, with the least distress as possible. Scientists conducting pre-phase clinical trials have to report to the government on such details as the size of the cages in which they keep animals, food provided, etc.

Provided there is an empirical basis (based on pre-phase trials) for thinking that a novel intervention might prove to be a safe and effective treatment for a disease or disorder, a phase 1 clinical trial can begin. Patients that agree to participate in a clinical trial will be made aware of all relevant information (e.g. what the intervention is, why this research is taking place, all the known side-effects/toxicity of the drug, etc.) and can either consent to participate or decide not to do so. A phase 1 study has a small number of participants – twenty to a hundred – either healthy volunteers or people with the disease or condition. In the United States, approximately 70 percent of drugs successfully pass the phase 1 stage, which can take several months to complete.

If this first stage is passed successfully, meaning no significant safety concerns arise, phase 2 trials then involve testing on larger numbers of participants (several hundred). In phase 2 the intervention is tested for safety and efficacy. This involves having a control group taking a placebo, specifically designed to have no real effect. This will help researchers determine the actual efficacy of the intervention being tested. Phase 2 trials can take anywhere from several months to two years. And approximately 33 percent of drugs pass this stage onto phase 3.

Phase 3 trials can take from a year to four years and involve 300 to 3,000 volunteers. The purpose of this stage is to further test the efficacy of the intervention as well as monitor for adverse side-effects. Approximately 25 to 30 percent of drugs in this phase successfully move to phase 4, which is carried out once the intervention has been approved by the FDA during post-market safety monitoring. Many clinical trials, at any of the phases, can be delayed because of a shortage of participants. Developing a safe and effective medical intervention is a lengthy, costly and potentially risky endeavour. A virtuous polity will seek to find a fair and feasible balance between the different interests and stakes that arise in this context, including the imperative to advance knowledge and medicine, the duty to protect the safety of the participants, and keeping the costs and administrative paperwork involved in check so that innovation can take place rather than be stifled. This is not an easy undertaking!

VII How virtue ethics can help us navigate the complexity and provisionality of the genetic revolution

This quick snapshot of the diverse advances being made in the field of genetics, and the complex ethical and social concerns they raise – from treating disorders more effectively to the worries of genetic discrimination and the regulation of intellectual property – effectively illustrates why some bioethicists have remarked that "coping with these new [genetic] powers will tax our wisdom to the utmost" (Buchanan et al. 2000: 1). Given the nature of the subject matter we are concerned with in this book – namely how to ethically advance genetic knowledge and potential interventions that might permit us to alter our genes and biology in novel ways – I believe a virtue-oriented normative lens is more appropriate than an approach that champions one particular principle or value. Appealing to a principle of benevolence might skew the moral lens by framing the issue as one of seeking only to help prevent the disadvantages (e.g. morbidity) caused by the genetic lottery of life. But when understood instead as a *virtue*, the moral landscape will be construed in a more nuanced fashion, so that we see that mitigating such harms will come with its own risks as well as costs and uncertainties. And navigating through these complexities is an integral part of the ethical enterprise.

Promoting knowledge and achieving medical breakthroughs are both important, but it is also important that participants in medical research provide their informed consent and that the science be rigorously tested for safety and efficacy. And, given the fact that there are always budgetary constraints on scientific research and healthcare provisions, difficult questions must be addressed concerning the extent to which it is better to pursue environmental interventions to promote equality and health rather than novel genetic interventions. No doubt a two-pronged strategy is required. A virtuous polity will keep an open mind about the prospects of pursuing more aggressive environmental measures (e.g. better air control, more healthy food options in schools for children, etc.) as well as new potential treatments such as gene therapy or gene editing.

Learning about the risks and costs of genetic interventions, suppose the critic says that all these considerations simply reinforce their conviction that we should not attempt to tamper with our genes in any fashion. "Don't try to play God!" they might contend. "Have the humility to recognize nature designed us the way we are for a reason." An appreciation of two facts about human life shows us why we should not embrace the "biological status quo" attitude of the critics of genetic engineering.

The first fact is the reality of the prevalence of disease, disability and suffering, especially in late life. Most humans over the age of sixty will experience multi-morbidity. And given that life expectancy at birth for the world now exceeds age seventy, and is expected to reach age eighty by the end of the century, the health challenges posed by an aging population raise significant concerns for both benevolence and justice – concerns that are not taken seriously if one simply sticks one's head in the sand and advocates for the "biological status quo" position.

Secondly, the biology we have today is not "the status quo" of our species. We have evolved, and will continue to evolve, over time. Whether as the result of a conscious and intentional intervention by us or not, the biology of our species will continue to be influenced by environmental factors and procreative decisions. We cannot stop this evolutionary process even if we wanted to. The real question is whether we should purposely seek to intervene to improve that process of evolution by natural selection or not.

In contrast to someone invoking a genetic "status quo" position, predicated either on precaution or on the claim that such interventions are "unnatural," imagine the overzealous enthusiast coming at things with a different disposition. Suppose, for example, they invoke a *principle of equality* to help us navigate through the ethical challenges of the genetic revolution. "Everyone should start life with the same biological potential for a healthy and happy life!" they laudably contend. This might lead one to propose that a principle of *genetic equality* is the ideal to strive for.

Elsewhere (Farrelly 2004)[8] I have noted the challenges with this principle. Firstly, the principle of genetic equality (GE) needs to address *the currency problem*. The currency problem requires us to answer the question "What should be equalized?" Should we try to equalize our genetic potential for health, height, intelligence, being physically attractive, etc.? All of these advantages are morally arbitrary, and thus, to be consistent with the logic of some egalitarian theories (i.e. luck egalitarianism),[9] egalitarians should endorse the view that any genetic potential that has a significant impact on our life prospects should, ideally, be equalized. Let us call this the *broad interpretation* of GE.

Of course there may be reasons why egalitarians would reject the broad interpretation of GE. They may recognize that such a proposal presupposes fantastical knowledge of how genes work and unrealistic assumptions about what our capabilities for genetic manipulation could ever be. Furthermore, this broad interpretation of GE, as Buchanan et al. (2000) point out, is susceptible to two other problems. Firstly, what counts as an asset is defined at least partly by the basic institutions and practices of a society. This means that the traits we view as valuable will inevitably change with time. If we intervene in the natural lottery of life to ensure that future generations have the same genetic potential to develop valuable traits, it may turn out that by the time they reach adulthood those traits will no longer be so valued. The traits valued in an agrarian society, for example, are vastly different from those valued in highly advanced industrial societies, where computer literacy is a prerequisite for a constantly growing number of occupations. We just do not have the foresight to be able to predict which traits will or will not be valued in the future. Thus, intervening in the natural lottery in the name of equality is bound to fail to achieve what egalitarians hope it will achieve (i.e. to make everyone equally advantaged).

A second problem with genetic equality, argue Buchanan and his colleagues, is that "any thought of equalizing [natural] assets would almost certainly betray a failure to appreciate what might be called the fact of value pluralism (or diversity of the good)" (2000: 80). There is no "objective list" of physical or behavioural characteristics that all reasonable people would agree are valuable, let alone the *most* valuable.

To avoid these difficulties, an egalitarian might endorse a more narrow interpretation of GE – one that does not apply to contentious traits. The *narrow interpretation* of GE maintains that all should be equal in terms of their genetic potential for goods such as health and longevity. Of course the narrow interpretation need not be limited just to these two goods; there may be other traits an egalitarian would like to include as part of the more narrow interpretation of GE. But this list would not include goods that would make the principle of GE vulnerable to the two objections raised by Buchanan and his collaborators. The narrow interpretation is attractive not only because it guards against those two particular objections, but because it is also sensitive to the fact that these technologies will be very costly (as we just saw above with the costs of medical research). The more we include within the range of genetic potential that should be equalized, the more public funds we need to invest in such technologies.

But prioritizing the investment of such funds in realizing "genetic equality" may not be wise, given that other factors, such as the environment, can sometimes play a more important role in influencing our life prospects. The more public funds we invest in equalizing our genetic endowments, the less we have for pursing equality in other important dimensions of society, such as equal opportunity for education, healthcare in general or socio-economic equality.

The issue of the *costs* of pursuing genetic interventions leads us to the second main problem that debates about genetics and justice need to take seriously – what I call *the problem of weight*. For egalitarians, this problem means balancing the desire to achieve genetic equality with the aspiration for promoting other kinds of equality (e.g. wealth and income) and other values (e.g. utility and freedom). The more expansive the interpretation of GE one defends, the greater the difficulty of resolving the problem of weight. This is so because, the more expansive the interpretation of GE, the more costly pursuing that principle is and, thus, the more difficult it will be for society to pursue other forms of equality.

The viability of the broad interpretation of GE is thus seriously strained once one takes the fact of scarcity seriously. Given the fact that environment plays such an important factor in influencing physical and behavioural characteristics, egalitarians will not be able to justify investing the amount of public funds needed to pursue GE when inequality in environmental influences mean that people will still end up unequal in terms of their education, income, attractiveness, etc. Rather than trying to equalize our genetic potential for physical and behavioural characteristics, egalitarians might decide that it is better to permit this inequality and pursue other forms of equality which will bring about more utility. While egalitarians believe that equality has considerable moral value in itself, it is not the only thing they value. Thus concerns for achieving GE must be informed by considerations of utility, and this will lead egalitarians in the direction of endorsing a more narrow interpretation of GE.

The narrow interpretation of GE is better suited to taking scarcity seriously than the broad interpretation, and thus it can combine concerns of equality with those of utility. However, both interpretations are ill-suited to shedding light on the weight we should place on the value of freedom. This is particularly important given the nature of genetic interventions. Egalitarians believe that some form of state coercion (e.g. taxation) is justified for achieving economic equality. But, given the nature of genetic interventions, egalitarians must take very seriously the issue of how we can *justly pursue* GE. While they will (rightly) dismiss the libertarian charge that taxation of income is a violation of self-ownership,[10] such concerns are much more pressing when it comes to the issue of pursuing genetic equality. GE cannot be achieved by taxing people's wealth; it requires genetic manipulation, and this could possibly conflict with reproductive freedom. The fact that such interventions might violate self-ownership does not necessarily mean they are unjust. But, given the atrocities of past eugenic movements, egalitarians must take seriously the value of freedom and ensure that it figures prominently in their account of genetics and justice. The principle of genetic equality does not provide any help in this regard, and this further limits the principle's appeal.

Rather than advocating the more grandiose aspiration of "genetic equality," the authors of *From Chance to Choice: Genetics and Justice* (Buchanan et al. 2000) argue that genetic intervention to prevent or ameliorate serious limitations as a result of disease is a requirement of justice. But even this more modest proposal can run into problems. What if the costs of pursuing the genetic therapy of an extremely rare but debilitating disease are so high that the only way a polity could fund it was by reducing (or even eliminating) the healthy meal options it provides to young children at public schools? In other words, how much weight should we place on bringing everyone past a genetic minimum threshold? Advocating a right to a genetic decent minimum (GDM) is a non-starter because rights cost money, and "nothing that costs money can be absolute" (Holmes and Sunstein 1999: 97). Taking scarcity seriously means taking seriously the fact that we must make tradeoffs in rights protection. A GDM, like GE, does not do this. Thus the practical import of a GDM is very limited. Such a principle might be of use in a society that already satisfied a decent minimum of other goods (e.g. housing, education, nutrition, wealth, etc.) *and* already possessed a vast supply of genetic therapies. But no society in this world is like that.

A GDM, like GE, also fails to take concerns of freedom seriously. To say that all should have a decent genetic constitution tells us nothing about how we should *pursue* this aim when, for example, the procreative liberty of a parent impedes our realizing such a goal. Current societies could already implement a GDM by adopting eugenic policies that regulate who can procreate with whom. A virtuous polity would reject such a proposal as it is a gross violation of freedom, but this example reinforces the point that fundamental distributive principles should be sensitive to the diverse pressing concerns that arise in the real world. A GDM, like GE, ignores many of these issues and thus is of limited practical import.

Rather than approach the topic of ethics and genetics with firmly held principled convictions, I think a virtue-oriented approach is more attractive and feasible. Such an approach, especially when it emphasizes both the moral *and* intellectual virtues, permits concerns such as safety and equality to play substantive roles in determining which courses of action, or which regulatory options, are most appropriate to adopt when we have imperfect or incomplete knowledge. As such it takes a much more nuanced and provisional stance than the position typically taken by a principled moral analysis. And I believe this is particularly important when it comes to the knowledge and technologies under consideration in this book.

Discussion questions

1 Advances in the biomedical sciences are progressing at an impressive pace. Do you believe it is important for scholars in the humanities and social sciences to ponder and debate the ethical and social implications of these advances before we determine how to regulate such innovations? Or do you believe scientists themselves are best positioned to resolve such normative concerns on their own?

2 According to a recent Pew Research Center analysis,[11] only 60

percent of Americans believe that "humans and other living things have evolved over time." Furthermore, only half (32 percent of the American public overall) of those who expressed a belief in human evolution take the view that this evolution is "due to natural processes such as natural selection." Approximately a quarter of adults say that "a supreme being guided the evolution of living things for the purpose of creating humans and other life in the form it exists today." What challenges does this divide between the insights of science and the opinions of the average person create for implementing sound science policy when it comes to advances in the biomedical sciences? Can you think of suggestions for narrowing this knowledge gap?

3 A just society aspires to prevent discrimination on the basis of race, sex, religion, sexual orientation, age, etc. The prospect of "genetic discrimination" thus raises new concerns for a virtuous polity. Genetic discrimination "occurs when people are treated differently by their employer or insurance company because they have a gene mutation that causes or increases the risk of an inherited disorder."[12] Do you believe it is important, and feasible, to protect people from genetic discrimination?

3
Disease

I Introduction

The Cancer Genome Atlas, a collaboration between the National Cancer Institute (NCI) and National Human Genome Research Institute (NHGRI), was launched in 2005.[1] Cancer is one of the leading causes of death worldwide, with an annual death toll of over 8 million people a year. The World Health Organization estimates that approximately one in six deaths is caused by cancer.[2] The goal of *The Cancer Genome Atlas* is to help reveal the biological basis of cancer. There are over 200 different types caused by errors in our DNA that thus result in uncontrolled cell growth.

To understand the social significance of projects such as *The Cancer Genome Atlas*, and the genetic revolution more generally, we must be cognizant of what medicine is and of what it aspires to realize. At the most basic level, medicine (and science more generally) is a *skill*. And all skills, argued Aristotle in the *Nicomachean Ethics*, strive for some ultimate goal or purpose (the Greek word is *"telos"*). What is the *telos* or purpose of medicine? Imagine, for example, Martians visited Earth to learn more about our civilizations. They were particularly intrigued by our medical practices. They observed activities such as the training of doctors and nurses, the questions physicians asked patients during their health checkups, the prescription of drugs for a wide variety of illnesses and diseases. They observed surgeries, ranging from minor surgeries to life-saving ones. After witnessing all of these things the Martians asked one simple question – *Why*? What is medicine for?

What would our general answer to this question be? The cynic might argue that most of modern medicine, at least in developed countries, is simply a lucrative way for pharmaceutical companies, and even physicians, to make money. There may be something to that critique worth taking seriously in terms of not over-idealizing the "status quo" of our current medical practices. But I think, as a response to the *telos* question, it is overly cynical, and, most importantly, mistaken. The best response, in my view, concerning the question of what medicine is for is this: *medicine aims to benefit the patient (typically the focus of clinical medicine) and a population (the focus of public health)*. We train physicians and nurses so they can care for the sick and educate their patients to help prevent harm (e.g. reduce the risk of disease). The drugs, hospitals and specialized branches of medicine (e.g. oncology, neuroscience, etc.) aspire to prevent, treat and manage a wide host of maladies. And public health measures such as immunizations and clean drinking water seek to promote the health of a population, to help protect members of the community from becoming patients in the first place. So medicine encompasses the knowledge, practices and technologies that a polity can exercise, deploy or provide to prevent, diagnose and treat or manage pathology, disability and suffering.

Medicine is thus a crucial element of a *benevolent*, as well as a *just* society. Whatever the failures of our current polities may be, those of us living in developed countries who enjoy access to public health measures and medical resources and expertise should be grateful for the knowledge, practices and technologies that help us flourish as individuals and as a polity. Medicine often *epitomizes* the exercise of the moral and intellectual virtues, even if our exercise of these virtues is often imperfect. In this chapter we shall consider the so-called detective's virtues in greater detail, relating them to the issues of disease, genetics and medicine.

II Why is there disease?

Let us return to the thought experiment of the Martians for a moment. Suppose the Martians asked a second basic question: *Why is there disease?* They ask this question because they have no diseases on their planet. No Martian ever dies from things like cancer, malaria or heart disease. Thus they are puzzled by the fact that millions of humans die from these diseases each year. They want to understand why the disease burden of our species on planet Earth is so high. How would we answer their question? Any comprehensive answer we provide should exemplify *the detective's virtues*. This means it should invoke coherent explanations of the facts. So perhaps we should begin with some facts about disease.

There are two general types of diseases – infectious and chronic. Infectious diseases, such as HIV/AIDS, malaria and smallpox, are spread through contact (e.g. unprotected sex), water, mosquitoes, etc. Chronic diseases include diseases such as cancer, heart disease and stroke. We will turn to the issue of what causes chronic disease shortly. But let us start with infectious disease.

III Infectious disease

Why are there infectious diseases? The reality is that our world (unlike our fictional Martian's planet) is a hostile place for all life forms, including humans. A total of 1,415 species of infectious organisms have been identified as causing disease in humans, of which 217 are viruses and prions, 538 bacteria and rickettsia, 307 fungi, 66 protozoa and 287 helminths (Taylor et al. 2001).

Smallpox was the scourge of the twentieth century, and "some estimates put the death toll caused by smallpox in the twentieth century alone at 500 million people" (Koplow 2003: 1). It is believed that smallpox originated approximately 3,000 years ago, though "unmistakable descriptions of smallpox did not appear until the 4th century AD in China, the 7th century in India and the Mediterranean, and the 10th century in south-western Asia" (Fenner et al. 1988: 210).

Smallpox is the only disease that has been eradicated (since 1980) worldwide through vaccination. Vaccines have helped bring other diseases under control. It took almost two centuries (and hundreds of millions of deaths) from the time when the English doctor Edward Jenner first demonstrated that inoculation with cowpox could protect against smallpox till the time when the disease was finally eradicated. The example of smallpox clearly illustrates how important it is for society to implement "well-ordered" science: "The pursuit of science in a society is well-ordered when the research effort is efficiently directed toward the questions that are most significant" (Flory and Kitcher 2004: 59). Had we developed and distributed the smallpox vaccine decades earlier, millions of lives could have been saved.

In addition to viruses, humans are vulnerable to hundreds of bacteria. Many bacteria cause *water-borne diseases*, such as cholera, E. coli and dysentery. Water-borne diseases, like infectious diseases more generally, are a much more significant problem for developing countries that still face the challenge of providing clean drinking water and sanitation.

According to the World Health Organization's *Progress on Sanitation and Drinking-Water: 2015 Update*,[3] 2.4 billion people still lack improved sanitation facilities and 946 million practice open defecation. And 663 million people still lack improved drinking water sources. So poverty and a lack of resources and technology is a major contributing factor to the risk of infectious disease. And because epidemic diseases are influenced by climate conditions, global warming can increase the risks. The rise in CO_2 emissions caused by humans since the Industrial Revolution has contributed to the warming of the Earth's surface temperature, and this can have an impact on things such as the prevalence of malaria (Tanser et al. 2003).

When explaining why infectious diseases plague humans on our planet, part of the story we would tell the Martians is the story of our inhospitable world – the extrinsic risks of the environments in which we live. Perhaps the Martian planet (keep in mind it is a fictional planet!) does not contain such hazards, hence why Martians do not develop disease or need medicine. When we explain the cause of disease by invoking the extrinsic risks of our world we are invoking what is called a *proximate* explanation. We explain why disease occurs by identifying immediate causal factors (such as the existence of viruses and bacteria) that lead to the development of specific diseases.

But what if the Martian planet also possessed the very same infectious organisms that we have here on Earth? And yet no Martian develops malaria, smallpox, cholera, HIV/AIDS, etc. What proximate-level explanation could explain this difference in morbidity risks? The answer would have to lie with differences in *the biology* of humans and of Martians. Perhaps the Martians have a different immune system, or blood cells, or metabolism, etc., that protect them from infectious diseases such as influenza, HIV/AIDS and malaria. The genes our Martians possess must be very different than those we have inherited from our evolutionary history. Thus an important part of the story of our vulnerability to infectious disease is the story of our *evolved biology*. Telling that element of the story requires us to invoke the *ultimate*, or evolutionary, explanation of why there is infectious disease.

Pathogens are arguably the strongest selective pressure to drive the evolution of modern humans (Karlsson et al. 2014: 390). To understand why this is so, it may be helpful to review the basics of evolution by natural selection. Charles Darwin's idea is one of the greatest scientific insights in human history. Evolution by natural selection occurs for a population when there is (1) *a variation in a trait* (e.g. height, resistance to local pathogens, length of a beak, aggression, etc.), (2) *differential reproduction* (e.g. some individuals have greater reproductive success than others) and (3) *heredity* (e.g. genes that influence height, immunity, a long beak, aggression, etc.). Over time advantageous traits become more common.

If particular individuals possess genes that help make them more resistant to local pathogens than others, it is not difficult to understand why the former would fare better with reproductive success: they are more likely to survive long enough (1) to reproduce and (2) to care for their offspring for a sufficient length of time for them also to reproduce. So the advantageous genes – that is, those that increase *reproductive fitness* – will become more prevalent over time.

The tragic example of the flu pandemic of 1918, which killed an estimated 50 million people worldwide (Watanabe et al. 2008), is a vivid illustration of this point. What made the 1918 virus particularly devastating was the fact that nearly half of the influenza-related deaths occurred among young adults (twenty to forty years of age) (Simonsen et al. 1998: 53). This was perplexing, as it is usually the very young or the elderly who are most at risk of death from influenza, as these vulnerable members of a polity have less immunity. By reconstructing the 1918 virus, researchers found that particular genes may have played an important role in the virus's ability to obstruct airway cells. HA, NA and PB1 genes, for example, have been identified as being those most optimal for virus replication and virulence (Pappas et al. 2008). "Understanding the molecular basis of the high-virulence phenotype of the 1918 pandemic virus is important, as it could identify useful targets for drug intervention when new pandemic viruses begin to emerge" (Watanabe et al. 2008: 590). Thus advances in our understanding of the role played by particular genes in our susceptibility to infectious diseases could lead to new drugs and medicines that could protect populations from SARS or H1N1, for example.

HIV/acquired immunodeficiency syndrome (AIDS) and malaria are two other infectious diseases where genetics are important. HIV can be transmitted through sexual contact (e.g. unprotected sex), through blood (e.g. blood transfusions or sharing a needle for drug use) or through mother-to-child transmission. Some individuals exposed to high-risk activities, for example commercial sex workers, have remained uninfected by HIV (Fowke et al. 1996). These (rare) individuals have a *natural immunity* to the disease that most of us do not enjoy. Possession of a CCL3L1 copy number lower than the population average is associated with markedly enhanced HIV/ AIDS susceptibility (Gonzalez et al. 2005). A better understanding of how particular genes might help protect against the risks of HIV/ AIDS could lead to new medicines such as a genetic vaccine or gene therapy for HIV.

Malaria is transmitted via bites from infected mosquitoes. As such, it is a tropical disease that typically occurs in temperate regions of the world. "Malaria has been a major determinant in the evolution of several human genes, especially those involved in the constitu- tion of red blood cells" (Guégan et al. 2008: 25). The higher risk that African Americans face with respect to sickle cell disease, for example, stems from this fact. Sickle cell disease affects approxi- mately 100,000 Americans and occurs in about one out of every 365 black or African-American births vs. one out of every 16,300 Hispanic-American births.[4] Individuals born with two copies of the HbS gene acquire sickle cell disease because they develop distorted red blood cells. And this reduced blood flow causes damage to tissues and can lead to death. But individuals born with one copy of the HbS gene and one copy of the "normal" gene variant do not typi- cally develop any symptoms. Furthermore, having just one copy of the abnormal gene confers "significant protection against all-cause mortality, severe malarial anaemia, and high-density parasitaemia" (Aidoo et al. 2002: 1311).

When we provide a detailed answer to our Martian's query "Why is there disease?," the detective's virtues remind us that a coherent explanation of infectious disease will emphasize both environmental risks (the existence of infectious organisms) *and* our own evolved biology (and its inherent vulnerabilities and limitations). Why do we have the vulnerabilities and limitations that we have? Why aren't we just immune to all possible infectious diseases (like our fictional Martians)? Why do humans have the lifespan we typically have? Why can we experience fear, sadness and anxiety instead of just constant happiness, curiosity and arousal? And why is there a variation among us with respect to all of these phenotypes? To answer these questions, the detective's virtues instruct us to give attention to insights from evolutionary biology.

IV The challenge of our times: chronic disease

Let us now turn to chronic disease. While the history of humanity before the twentieth century was the history of a world dominated by infectious disease, the world of the twenty-first century is one (barring a new worldwide outbreak of an infectious disease) dominated by chronic disease. Important facts to emphasize are (1) the magnitude of the chronic disease burden and (2) the proximate and evolutionary causation of chronic disease.

In the decade from 2005 to 2015, the World Health Organization estimates that approximately 76 million people died from chronic illness in high-income countries.[5] But the number is even higher for the more populous lower-middle-income countries, such as China and India. It is estimated that, in that same ten-year period, chronic illness caused 144 million deaths in these countries.[6] "As risk of death from infectious and parasitic diseases diminish, the degenerative diseases associated with ageing, such as heart disease, stroke and cancer, become much more important" (Olshansky et al. 1993: 47).

Like the story of our risk from infectious disease, the story of our risk to chronic disease is one about our environment (e.g. diet, tobacco, sedentary lifestyle, pollution, etc.) and our biology (especially aging, or *senescence*). Smoking can cause lung cancer, a diet high in sodium can lead to hypertension, a sedentary lifestyle can increase the risk of cardiovascular disease, etc. During their observation of physicians meeting their patients in developed countries, the Martians would have heard the doctors prescribing the health benefits of a healthy diet and lifestyle. If such advice were heeded, this could help delay the onset of many chronic diseases.

But the extrinsic risks of the food we eat, the air we breathe, and the lifestyle we live is only part of the story of our risk of chronic disease. The vast majority of people who die today from cancer, heart disease, stroke, etc., die in *late life*. It would be a major omission to leave this out of our explanation to our disease-free inquisitive Martians as to why there is disease. "As people get older, we become more susceptible to disease," we tell the Martians. But they again look perplexed. They explain that, for their species, the number of years a Martian has already lived has no bearing at all on their risk of disease. They are *biologically immortal*, meaning they die only if there is an accident – for example a spaceship exploding – or they fight each other, or they don't eat, etc. But their minds and bodies don't fail (e.g. they don't suffer age-related arthritis, diminishing bone density, dementia, etc.) simply because they are chronologically older. Like the earthly hydra, our Martians have regenerative capabilities that protect them from the harms of senescence. Why aren't humans like the hydra? Why do we, and sexually reproducing species in general, age *biologically*, and thus become more susceptible to chronic disease and death?

Scientists have an answer to this question, and the answer is provided by the *disposable soma theory*.[7] This account maintains that biological aging occurs because natural selection favours a strategy in which reproduction is made a higher biological priority (in terms of the utilization of resources) than the somatic maintenance needed for indefinite survival. The story of the intrinsic constraints of our biology thus begins with the story of the world's extrinsic risks. Bruce Carnes (2007) usefully describes this theory as follows. The world is a dangerous place. Death is, for all living things on this planet, inevitable. In order for any species' existence to persist over time a solution to death must be found. And that solution, for us and for other sexually reproducing species, is reproduction. There is thus a real race between reproduction and death, and all the species alive today are, at least for the moment, winning this race. But for all the species that are now extinct, such as the mammoth and Neanderthals, the race was lost.

Adopting this organismal perspective is important because it brings to the fore the fact that a biological tradeoff must be, and has been, made between the physiological resources we invest in reproduction and those invested in the maintenance of the soma. Whatever the evolutionary history of our biologically immortal Martians is, they must have found a different solution to death (considering they are still alive!) than the strategy pursued by sexually reproducing species on Earth. For us, reproduction is given a higher biological priority, which means that, while our bodies and minds are not designed to fail, a consequence of the priority placed on reproduction is that there are inevitable health problems as the "biological warranty period"[8] expires (typically by the seventh decade of life).

To push the Martian example just one more fanciful step forward, I request the patience of my reader in granting one more unrealistic assumption. Suppose our Martians tell us that, in addition to living on a planet that has many of the pathogens that we possess, their evolutionary history was also one that adopted a similar reproduction/longevity tradeoff. And for millennia they too suffered from age-related diseases such as cancer, heart disease and stroke. But eventually they developed new medical innovations: first, an intervention that slowed down the aging process (thus delaying the onset of chronic disease) and then, eventually, technologies that eliminated aging altogether. In this scenario, our last twist in the thought experiment, the reason there is a difference in the morbidity risks between us and our visiting Martians is not the absence of pathogens in their environment, nor is it a difference in evolutionary history or genetic makeup. The main difference is one of *knowledge*. The Martians possess something that we currently do not – the knowledge and technological ability to retard, even eliminate, the aging process.

Putting together all the elements of our fable of the Martian's query "Why is there disease?," we can see different lenses that help explain the existence, and persistence, of disease in the world today. One lens illuminates the existence of the *external threats* of our world – infectious organisms, and another illuminates *internal features of our biology*, such as the presence of *particular genes* or *evolved biological tradeoffs*. Some genes can be beneficial (insulating some from HIV or helping to protect against the diseases of late life) and others detrimental (causing single-gene disorders).

And a third and final lens is our *level of ignorance* vs. *knowledge and technology*. One reason infectious diseases ravaged human populations for most of our evolutionary history was that we were ignorant about these infectious organisms. We could not observe, with the naked eye, viruses and bacteria. Unaware of their existence, we were ill-prepared to do anything effective to protect human populations from things such as cholera or smallpox. But over the past two hundred years humans have gained new insights into the risks of disease, and this knowledge has led to major public health and medical advances, such as sanitation, immunizations, safer food, birth control, new surgical techniques, drugs to treat and manage multi-morbidity, etc. So *epistemic vice* (e.g. ignorance, ignoring the facts, etc.) has proved immensely costly to humans in terms of the lives lost prematurely to disease.

Our environment, our biology and *our knowledge* are all extremely important elements to emphasize in explaining the disease and health prospects of human populations. Advances in our knowledge (for example epidemiology) have helped us to shape safer environments and alter our biology by boosting our immunity through vaccinations. Changes in diet and lifestyle can reduce the risks of chronic disease. And it may be possible to deliberately alter specific genetic mutations to prevent or treat specific diseases.

Once these elements of the detective's virtues are emphasized, we can understand why a virtuous polity would avoid the folly of both environmental and genetic *determinism*. Genetic determinism naively assumes that the genes we inherit determine particular phenotypes such as disease, health or behaviour. But the environment clearly influences these things as well. Being exposed to toxins, not wearing a seat belt, having parents with a secure attachment style, etc. – all of these things influence our biology and life prospects. Conversely, environmental determinism is clearly wrong. "According to environmental determinists, our educational and nutritional environments fully determine who we are, while genes make no contribution beyond perhaps fixing our membership of the human species" (Agar 2004: 71). We know now that heredity is very important not only for our health prospects but also for intelligence, happiness and behaviour. Genes and environment are both very important, as is the interaction between them (which we explore more fully in the next chapter, on epigenetics).

To place all our faith in the aspiration just to create healthy environments would be to overlook and ignore the fact that our biology itself can be a significant risk factor that we may be able to modulate safely (as in the case of vaccines). This is particularly important given the fact that the world's populations are aging, and thus the prevalence of the chronic diseases of aging are projected to continue to rise this century. Conversely, to place all our faith in the hope of developing new genetic therapies and interventions while ignoring the potential impact of poverty and climate change on the health prospects of those living today would be a mistake. A virtuous polity ought to continue to develop the knowledge and technology that would permit safe and effective health interventions to deal with both the external world and our internal biology.

The moral virtues of benevolence and justice instruct us to eschew any "status quo bias" in favour of some "natural" environment or biology. The "natural" environment of this world is extremely hostile to human life. That is why life expectancy at birth for humans did not exceed the age of thirty for most of our evolutionary history. It is an amazing success story that we live in a world where the average life expectancy at birth now exceeds age seventy.

And the idea that there is a "natural" biology of humans is illusory given the fact that it is constantly, though slowly, changing and evolving. The biology we possess today, including our genes, has been shaped by our environments *and* our culture. We possess a biology that has evolved through the process of natural selection. It is not the product of a grand omnipresent designer conceived to keep us healthy and happy indefinitely. This is important to recognize, for it helps us think more rationally and cogently about the question "Should we aspire to *consciously and purposively* modulate our biology through, for example, genetic intervention?" To answer "yes" is to recommend moving from unintentional genetic modification (UGM) towards intentional genetic modification (IGM) (Powell and Buchanan 2011).

If our Martians revealed to us a host of safe and effective genetic interventions – such as gene therapy and genome editing – that would help us eliminate cancer, heart disease and stroke, would we graciously accept those medical interventions? Or would we tell them we prefer to maintain our "disease status quo," despite spending billions of dollars every year on medical research on specific diseases? Would we tell them, invoking Michael Sandel's concern in *The Case against Perfection*, that such technologies threaten to banish our appreciation of life as a gift?

It is hard to see why any virtuous polity would not accept these medical interventions if they were indeed reasonably safe and effective.

Consider, for example, if the intervention offered was just the modest advance of cheap and effective personal genome-sequencing technologies. This would not alter any person's genes, but it would allow us to tailor specific treatments to the patient more effectively based on their unique genetic profile. This could help avoid serious adverse side-effects from medications as well as improve the efficacy of medicine. What might appear as even a modest advance in our knowledge could be enormously beneficial, permitting us to shift from a medical model designed "for the general population" to a model tailored "to each specific patient."

Or imagine (more ambitiously) the Martians offered us gene therapy for lung cancer or genome-editing technologies that could eliminate Alzheimer's disease. What if they had developed an "anti-aging" pill that could reduce the prevalence of age-related diseases by activating the "longevity genes" that permit humans to enjoy many more decades of healthy life? Would we welcome their gifts or would we reject them?

Recall what we identified as the *telos* of medicine: *medicine aims to benefit the patient (typically the focus of clinical medicine) and a population (the focus of public health)*. A virtuous polity would be foolish to emotively reject the medical benefits offered by the Martians if such knowledge and technologies would truly improve human health. But of course in the real world, unlike in a philosopher's thought experiment, the benefits of the genetic revolution will not fall like manna from the heavens. They will, like most medical interventions, come with associated risks and costs. They might, at least initially, be available only to some (e.g. the affluent) but not to everyone. So a virtue ethics analysis of the genetic revolution must consider the real-world complexities, and not just our idealized Martian example, if we hope to invoke the idea of a virtuous polity in a helpful fashion. In chapter 6 we will consider a host of concerns and objections typically raised against the prospect of extending the human lifespan by altering the aging process.

V Moving beyond "negative biology"

The virtuous polity which exemplifies the "intellectual virtues" would not limit the medical sciences to the study of just pathology.

Benevolence instructs us to pursue knowledge and innovations that could benefit others. And there are many instances of health research that could be beneficial which do not involve studying or trying to cure or prevent specific diseases. Elsewhere[9] I draw a distinction between what I have called "negative biology" and "positive biology." The vast majority of medical research conducted today falls under the paradigm of "negative biology" – that is, most researchers use an intellectual framework that presumes that the most important question to answer is *What causes pathology?* Disease is the central focus, and this explains why medical research and research funding is primarily concerned with trying to understand, prevent and treat specific diseases. The design of the US National Institutes of Health (NIH), which is largely composed of individual institutes dedicated to specific diseases, reflects this prevalence of pathology-oriented negative biology.

The NIH is the largest source of funding for medical research in the world. With an annual budget (in 2017) of $33.1 billion, it awards thousands of competitive grants to researchers working in the USA and around the world. The NIH is composed of twenty-seven distinct institutes and centers, among which are the National Cancer Institute (NCI), the National Institute on Alcohol Abuse and Alcoholism (NIAAA), the National Institute of Allergy and Infectious Diseases (NIAID) and the National Institute of Neurological Disorders and Stroke (NINDS), just to name a few.

With so many institutes dedicated to the study of specific diseases and disorders, one might draw the conclusion that well-ordered science and medicine mandates that the bulk of our energies and resources ought to be invested in trying to answer the fundamental question "What causes pathology?" But the virtue of intellectual humility cautions us against being too zealous in our convictions that we know which questions are the most important ones to ask.

Supplementing negative biology with what I refer to as the paradigm of "positive biology" helps provide a virtuous polity with a more intellectually curious and adaptive framework, one which could yield important knowledge to help realize the virtues of benevolence and justice. Recall, from a point noted in the introductory chapter, *"there is no single 'normal' human genome sequence; rather there are multiple variant human genomic sequences"* (NASEM 2017: 106). Why would it be virtuous to limit medical research to examining just "disease genetics" when there also exist the genetic puzzles of people who can remain disease-free over a century, or people with exceptional intelligence, exceptional emotional resilience, etc.? Intellectual virtue encourages, indeed compels, us to strive to unravel these biological puzzles and search for ways to improve the opportunities humans have for health, intelligence, happiness, etc. Focusing exclusively on the biological puzzles of pathology contravenes the virtues of humility and adaptability of intellect.

Rather than making pathology and disease the central focus of intellectual efforts and financial investments, positive biology seeks to understand positive phenotypes: Why do some individuals live more than a century without ever suffering from the common chronic diseases that afflict most humans much earlier in their lives? Why are some individuals more happy, more optimistic, more talented, or have a better memory than the average person? The paradigm of positive biology is based on the insight that the process of evolution by natural selection does not create a perfect organism in terms of lifespan, resistance to disease or other abilities.

Observations of exceptional longevity or superior cognition or resilience present fascinating puzzles for positive biology: Which biological processes (genetic, epigenetic, neural, etc.) explain these exemplars of health and wellbeing? The goal of understanding positive phenotypes is that such knowledge might lead to new interventions that could improve our ability to flourish as both individuals and polities. This might be achieved by modulating the rate of aging or by increasing opportunities for play and joy at all stages of the human lifespan, or by developing pharmaceuticals that safely enhance cognition or positive emotions, and so on. This is distinct from negative biology, which typically focuses on the proximate causes of specific diseases rather than on the evolutionary causes of positive phenotypes. Negative biology presumes that health, survival and happiness are the default states and thus aims to explain the deviations, such as Why do we develop cancer? Why do we suffer from depression? Why do we develop hypertension? Negative biology therefore faces the laudable, but overwhelming, task of trying to prevent or cure *all* disease. This is a costly and arguably ultimately insurmountable endeavour.

Eliminating all types of cancer would increase life expectancy at birth in the USA by 3.17 years for females and 3.2 years for males (Olshansky et al. 1990). Even eliminating cancer as a cause of death would not prevent any of the other chronic diseases of aging – cardiovascular disease, Alzheimer's, diabetes, and so on – from afflicting the elderly. Moreover, nearly half a century ago US President Richard Nixon declared a "war against cancer," and yet not a single type of cancer has been defeated. We still have a long way to go before we can realistically expect to reap the three-year increase in life expectancy that eliminating all cancers could yield. In fact, negative biology has not yet developed a single cure for any one of the numerous chronic diseases that afflict millions of people living today.

Why is medical research dominated by the study of pathology? A complete explanation would no doubt identify many distinct factors, but I believe that two important ones are worth noting which can help us appreciate why we need to go beyond the aspirations of negative biology. The first factor is that humans are susceptible to "observation biases." We are more likely to observe human tragedy and suffering than we are human health and happiness. When driving on the highway we slow down to closely observe car accidents, but we seldom note the fact that most drivers on the road avoid accidents and arrive safely at their destinations. The image of the carnage of a car accident can have a moving and lasting impression upon our psyche; the same is not true of the observation of the absence of accidents. The same cognitive bias occurs in our thinking about human health and our biology. It is natural for us to want to harness science to cure, treat and prevent specific diseases because these diseases are what kill most humans living in the world today.

A second reason we have invested so much in the focus on specific diseases is that this approach reaped enormous health dividends in the twentieth century by reducing early and mid-life mortality. Over the course of that century, life expectancy in the USA rose from forty-nine to seventy-seven years (Arias 2002: 29). This dramatic rise was caused by many factors, such as technological advances, increased material prosperity, changes in behaviour (e.g. birth control), etc., but the most significant medical advances arose from the application of insights from epidemiology's focus on the proximate causes of disease. This led to significant public health measures, for example the sanitation revolution, antibiotics and the smallpox vaccine. Focusing on pathology made a great deal of sense when the main cause of death was communicable disease. But the health challenges facing today's aging populations are different from those faced by populations in the early twentieth century. Expanding health in late life is far more difficult because of the reality of co-morbidity (Butler et al. 2008). The fixation on research for specific diseases has not resulted in the elimination of a single chronic disease. The pathology-focused strategy has helped increase the amount of time individuals in late life can be kept alive by managing multiple pathologies, but it has not made substantive improvements to a healthy lifespan.

To fixate our intellectual energies only on the study of pathology would mean that other important avenues of research would be neglected. A better understanding of exemplars of health and happiness – the goal of positive biology – might create more benefits for humans more quickly and more easily than could be achieved by negative biology alone. A drug that could safely mimic the effects of calorie restriction, for instance, might delay, simultaneously, most diseases and afflictions of aging. It would generate a much greater health dividend than defeating any one specific disease of aging, because slowing down the rate of ageing by seven years would reduce the age-specific risk of death, frailty and disability by about half at every age (Olshansky et al. 2006).

A lot of pioneering work in positive biology has already been undertaken in the burgeoning field of "positive psychology." Rather than studying why people suffer from mental illnesses such as depression, schizophrenia or ADHD (attention deficit hyperactivity disorder), positive psychology is primarily interested in how to improve the happiness of the "average" person. The things often assumed to make humans much happier, such as wealth, do not in fact have such a dramatic impact. Most disposable income is spent on consumer goods that do little to actually enhance our wellbeing.

In a study of the daily behaviour of happy people, researchers used an electronically activated recorder to document and then later classify participants' daily conversations with others as either "small talk" – that is, banal conversations – or "substantive talk" – where meaningful information was exchanged. They found that higher wellbeing was associated with less small talk and more substantive conversation (Mehl et al. 2010). While such a study does not establish the truth of Socrates' famous claim that "the unexamined life is not worth living," it does suggest that our need to feel attached to something larger is important to our happiness and wellbeing.

The psychologist Barbara Fredrickson's research (1998) on positive emotions – joy, serenity and gratitude – suggests that these expand cognition and behavioural tendencies. And finally, research on exemplars of resilience – that is, the ability of some people to cope and manage with tragic and traumatic events – could lead to the development of drugs that would increase resilience. Avshalom Caspi and his colleagues (2003) found that individuals with one or two copies of the short allele of the promoter of the 5-HTT serotonin receptor experience more depressive symptoms, diagnosable depression and suicidal thoughts in response to stressful events compared with individuals who are homo zygous for the long allele. In chapter 7 I shall consider the prospects of enhancing the emotional resilience of soldiers who will likely face significant trauma on the battlefield and the ethical concerns that this raises.

Cognitive functioning is another central topic of concern for positive biology. What are the genetic and environmental determinants of high IQ, exceptional memory and social intelligence? Researchers found that the analeptic drug modafinil significantly enhanced performance tests of digit span, visual pattern recognition memory, spatial planning and stop-signal reaction time in healthy volunteers (Turner et al. 2003). Improving average human intelligence could have a significant impact on our ability to overcome many of the world's most pressing problems.

Positive biology is not contrary to the goals and aspirations of negative biology. Indeed the two paradigms are often complementary. For example, understanding why some high-risk individuals (such as sex workers) seem to have an intrinsic resistance to HIV might spur the development of an HIV vaccine. Similarly, understanding human brains with exceptional cognitive functioning might lead to new avenues for developing drugs and therapies to prevent or treat severe cognitive impairment. Understanding exemplars of health could create real benefits for those who are more vulnerable to disease and disability. These findings of positive biology will eventually give us a better understanding of our human nature than the limited focus on pathology, and as such this might then lead to new genetic or environmental interventions or attitudinal changes and habits of mind that improve human wellbeing and happiness. Expanding the lines of enquiry of negative biology to include those of positive biology would help legitimate research into the determinants of happiness and wellbeing as integral elements of well-ordered science in the twenty-first century.

The contrast between negative and positive biology can be summarized as follows.

1 *Starting intellectual assumptions*
  Negative biology: Health, longevity and happiness are assumed to be a "given," or part of "normal species functioning," for humans.
  Positive biology: There is diverse variation in the genotypes which influence desired phenotypes such as health. The evolutionary and life history of different species helps explain this variation and the different biological tradeoffs that determine age of reproduction, body size, senescence, complexity of the brain, etc.
2 *What needs to be explained?*
  Negative biology: The proximate causes of disease, frailty and disability.
  Positive biology: The proximate and ultimate causes of exceptional health, positive emotions and happiness, high cognitive ability, etc.
3 *Which kinds of interventions ought to be pursued?*
  Negative biology: Interventions that help prevent, treat and cure specific diseases.
  Positive biology: Interventions that increase the opportunities for health, happiness and wellbeing.

Discussion questions

1 We can adopt three distinct lenses to explain the risk of disease for an individual and a population. One lens illuminates the *external threats of our world* – such as infectious organisms. A second lens illuminates *internal features of our biology*, such as the presence of particular genes or evolved biological tradeoffs. And the third lens is our *level of ignorance vs. knowledge and technology* to understand and mitigate the first two types of risk. Do you think we are prone to focus on just one or two of these explanatory perspectives, while ignoring or undervaluing others? Can you think of examples that illustrate the importance of all three lenses?
2 Chronic diseases such as cancer, heart disease and stroke have likely impacted your own life in important ways. Perhaps you have family,

friends or loved ones whose lives have been affected. Do you believe
the virtues of benevolence and justice require a virtuous polity to aspire
to prevent, as well as treat, chronic disease?

3 "Negative biology" studies the causes of morbidity and aspires to
   develop interventions that help prevent, treat and cure specific diseases. By contrast, "positive biology" studies the causation of exem-
   plar phenotypes, such as exceptional health, longevity, intelligence
   and happiness. Positive biology strives to develop interventions that
   can increase the opportunities for health, happiness and wellbeing. Are
   these both legitimate areas of scientific research for the medical sciences, or do you feel that negative biology should remain the central,
   perhaps even the sole purpose of such research?

4

Epigenetics

I Introduction

In earlier chapters we explored some of the ethical considerations that arise through purposively altering (or aspiring to alter) specific genes, or editing the genome of humans, in order to treat single-gene disorders or multi-factorial conditions. Insights from "epigenetics" also have important implications for the moral virtues, as they help explain the different ways *environment* can modulate our biology even if it does not involve altering the DNA sequence. "The origin of the term 'epigenetic' is often attributed to geneticist and philosopher Conrad Hal Waddington, who in 1942 used it to designate the 'branch of biology which studies the causal interactions between genes and their products, which bring the phenotype into being'" (Dupras et al. 2014: 330).[1] Cells can use DNA methylation, for example, to control gene expression. "DNA methylation regulates gene expression by recruiting proteins involved in gene repression or by inhibiting the binding of transcription factor(s) to DNA" (Moore et al. 2013: 23). DNA methylation is just one of several epigenetic mechanisms. In this chapter we shall consider the findings of different studies on poverty, the Holocaust, addiction and obesity, as well as paternal epigenetic inheritance. Each raises interesting questions about how a virtuous polity could potentially utilize such knowledge to better realize the demands of benevolence and justice.

The "epi" in "epigenetics" is Greek for "above" or "beyond," and it refers to "any process that alters gene activity without changing the DNA sequence, and leads to modifications that can be transmitted to daughter cells (although experiments show that some epigenetic changes can be reversed)" (Weinhold 2006, A163). The term "epigenetics" is often utilized in different ways and thus is something of a contested concept. But the definition provided above should be sufficient for the purposes of an ethical analysis of the issues canvassed in this chapter.

The reason it is important to address epigenetics is that these findings reveal how important the environment can be in altering our biology. A benevolent and just polity would not only pursue safe and effective interventions that directly alter genes and/or the genome, such as gene therapy or genome editing, to create more health and help prevent disease. It would also aspire to modulate *environments and cultures* to be more conducive to human flourishing. Virtue requires we pursue the mean between genetic and environmental determinism. Thus insights from epigenetics are an important piece of the puzzle in determining which courses of action are best to pursue to realize the demands of the moral virtues.

II Poverty

Theories of distributive justice have typically focused on the distribution of external goods such as wealth and income. In *A Theory of Justice*, John Rawls, for example, developed the most influential theory of social justice in the twentieth century, in which he maintained that socio-economic inequalities should be arranged so that they are to the greatest benefit of the least advantaged. Society's central institutions, for example the Constitution and political economy, Rawls agued, should seek to fairly distribute what he called the "social primary goods" – rights and liberties, powers and opportunities, income and wealth, and self-respect.

And a fair distribution of wealth and income, according to Rawls, was one that *maximized* the socio-economic prospects of the least advantaged, provided doing so did not violate higher-level principles of justice, such as fair equality of opportunity and the equal basic liberties principle. Rawls's original theory was developed in the early 1970s, and thus he did not take seriously the distribution of what he calls the "natural primary goods" – health, intelligence, imagination and vigour. One of his most prominent students, Norman Daniels, wrote his PhD dissertation (later turned into a book titled *Just Health Care*) expanding the Rawlsian account of justice to include healthcare among the things to be distributed in accordance with "fair equality of opportunity."

Many authors (Resnik 1997; Buchanan 1995; Buchanan et al. 2000; Allhoff 2005; Brown 2001), myself included (Farrelly 2016), have explored how the Rawlsian project might be extended to include principles that apply to the distribution of genetic endowments. But insights from epigenetics reveal that, even if we keep our focus just on the distribution of social primary goods, the impact environment has on genes further strengthens the demands of justice as they pertain to the distribution of things such as wealth and education.

Consider, for example, poverty. The virtue of benevolence prescribes that a society redress the vulnerabilities of poverty. Humans should not have to live under the duress caused by malnourishment or possible starvation. But the reasons for redressing such vulnerabilities extend far beyond simply mitigating the pain and suffering of starvation and death from lack of food. Poverty also profoundly impacts the biological development of humans, from our potential height and weight to our cognition and risk of disease in later life. Insights from epigenetics can thus further clarify, as well as potentially strengthen the stringency of, the moral demands of benevolence and justice.

For example, the potential epigenetics offers us in terms of providing further insights into the demands of justice and benevolence can be illustrated by considering the impact lower socio-economic status has on adolescents that have a positive family history for depression. If the risk of depression was purely genetic, one would expect to find roughly equal instances among adolescents from low- and high-income families. But this is not the case. Stress factors such as parental neglect and family discord can alter the expression of genes associated with depression, thus increasing the risks faced by an adolescent.

Depression is a medical condition that impedes a person's ability to live a flourishing life. An individual suffering depression will experience feelings of severe despair over a prolonged period of time. This might cause them to have difficulty concentrating, feel fatigued and hopeless, experience sleep disruption, and have thoughts of suicide. In some cases depression can result in suicide. The World Health

Organization estimates that, globally, 300 million people suffer from depression and that depression is the leading cause of disability.[2]

Virtuous polities would seek to prevent and treat depression on grounds of both benevolence and justice, and the strategies invoked to realize such aims should be diverse and varied. The virtue of benevolence instructs us to prevent harm and suffering when we can. The harms of depression are unequally dispersed upon individuals in a population, affecting the rich and poor, males (e.g. higher suicide rates) and females (e.g. post-partum depression), in different ways over the course of the lifespan. New insights from the epigenetics of poverty can further strengthen the conviction that benevolence and justice require us to redress the societal problems of poverty and economic inequality.

Pursuing, for example, only potential genetic interventions and a "genetic determinism" mindset to prevent depression is unvirtuous because, "although a positive family history is one of the strongest predictors of the future development of depression, not all individuals with this risk factor will become depressed" (Swartz et al. 2017: 210). For example, in "An epigenetic mechanism links socioeconomic status to changes in depression-related brain function in high-risk adolescents," Swartz and her colleagues examined the methylation of gene regulatory regions associated with lower socio-economic status. And this study found that lower socio-economic status was associated with a greater risk of depression among adolescents that had a family history of depressive symptoms. In other words, even if two people inherit the genes associated with higher risk of depression, if one of them lives in an environment with more adversity and stress, that individual will be at a higher risk than the other individual with the same genetic risks. Epigenetics reveals the important link between environment and gene expression.

From the standpoint of one prominent account of distributive justice, known as "luck egalitarianism," the fact that some individuals possess less wealth and income than others is an injustice if such inequalities are the result of chance rather than choice. That is, if two persons have unequal amounts of wealth because one person happened to be born into a family that left her a large inheritance, whereas the other person was born into a family with no such family wealth, this would be an *unchosen* inequality. Similarly, if one person was born with natural talents likely to command a higher income than the natural talents of another person, this again would be an inequality that is the result of *chance*. Luck egalitarians maintain that such inequalities should be mitigated as a matter of justice.

Insights from epigenetics permit us to develop a much more nuanced account of why such inequalities might be morally objectionable and in need of redressing than what is offered by luck egalitarianism's "conceptual-level" account of inequality. The problem isn't simply that some people have more money than others, but that lower socio-economic status can adversely impact important biological mechanisms that influence our potential to flourish.

In their study, Swartz et al. (2017) found that lower socio-economic status was a stressor that could increase the risk for future depression in high-risk adolescents (namely those with a family history of depression). Adversity such as parental neglect and family discord, and environmental risk factors such as poor nutrition and smoking, can impact developmental changes in DNA methylation. The authors conclude that the risk pathway they identify "could represent a discrete biomarker that could be targeted by novel strategies for personalized treatment and prevention" (ibid.: 213). That is, if adolescents with risk-related brain function can be identified *before* they become depressed, they could receive training in mindfulness-based techniques which might lower threat-related amygdala reactivity. Insights from epigenetics can further illuminate what is entailed by the virtue of justice.

The provision of decent public education – that is, the provision of intellectually, emotionally and physically engaging, nurturing and supportive learning environments – is one way the virtuous polity can provide a baseline of a positive environment for all children and adolescents to help mitigate the influence of socio-economic disparities and differences in the formative influences of home life. So quality, accessible education ought to remain a major societal priority for a virtuous polity. No advance in genetic screening or genome editing will displace the importance of education given the fact that cognition and health are dramatically shaped by both *environment* and genes. A virtuous polity will avoid the extremes of genetic and environmental determinism.

A further major study that reveals the importance of insights from epigenetics, especially for intergenerational justice, is the study on the association of the preconception trauma of the Holocaust on the stress levels of future generations (Yehuda et al. 2016). This means that the gross injustice of the Holocaust had effects that impacted not only those living during the Nazi regime; it also impacts the biology of those born *after* such atrocities have taken place. The descendants of the survivors of the Holocaust can suffer adverse effects from the stress of such events. These descendants have different hormone profiles (low levels of cortisol) than the regular population, profiles that make them more susceptible to anxiety disorders. If traumatic events such as the Holocaust impact epigenetic heritage, it is not unreasonable to think that there might be intergenerational epigenetic effects from other stressful events, such as slavery or the Indian residential school system in Canada (Bombay et al. 2014), where 150,000 indigenous children were removed and separated from their families and communities to attend residential schools in which sexual and physical assaults took place. New insights from epigenetics might help shed light on the demands of justice and benevolence that societies must address as a result of past events that adversely affected the biology of different populations.

III Nature *and* nurture: the case of addiction

The virtuous polity, I have argued throughout this book, would avoid the extremes of both *genetic determinism* – the belief that health and behaviour are determined solely by our genes – and *environmental determinism* – the belief that health and behaviour are determined solely by environment. Rather than believing that it is only nature or nurture that shapes our biology, the virtuous polity aspires to understand the complex reality that both nature and nurture are important. This isn't to deny that there are clear cases where genes or environment profoundly influence, indeed determine, the development of a phenotype. For example, if you develop the genetic mutation for a single-gene disorder such as Fragile X syndrome, then you will develop Fragile X, which affects a person's behaviour and intellectual development. The impact of Fragile X on intellectual functioning can range from mild (e.g. learning disorders) to severe (e.g. intellectual disability). There are clearly cases where specific genetic mutations profoundly shape a person's life prospects, as in the case of single-gene disorders. Conversely, there are clearly cases where the environment in which a person lives can profoundly influence their life prospects. If you are born into a country that lacks clean drinking water or nutritious food or suffers from recurring civil unrest or war, your chances of living a long and flourishing life are dramatically reduced. But, outside of these obvious cases, many forms of disadvantage (developing cancer, depression, addiction, etc.) are influenced by both genes and environment. And thus the findings of epigenetics are essential to ensure we avoid the mistakes to which we could be susceptible if we subscribed to either genetic or environmental determinism.

In "Epigenetics and the environment in bioethics," Charles Dupras et al. (2014) argue that individualist and rights-based forms of bioethics have neglected the importance of the environment in determining individual and public health. By making the individual patient the central focus of normative analysis, the authors maintain that North-American bioethics distorts the moral landscape by focusing an intense moral lens on autonomy and personal rights but neglects the importance that environment has on individual and public health. The virtue ethics lens I develop in this book is very sympathetic to the complaints Dupras and his colleagues have against mainstream North-American bioethics. By emphasizing the moral and intellectual virtues, I have attempted to steer my normative analysis in a direction that engages with important insights from public health and evolutionary biology, as well as clinical medicine. And expanding the scope of the analysis of genetics and ethics to include a concern for the role environment plays in human health and behaviour also helps to ensure a more balanced moral analysis.

The importance of environment can be further illustrated by briefly examining the case of addiction. Addiction is an effective example to consider because it easily reveals that the problem cannot be looked at through the dichotomy of "nature vs. nuture"; rather, we need to understand the specifics of how both "nature *and* nurture" influence the development of different types of addictions.

"Addictions are psychiatric disorders that are associated with maladaptive and destructive behaviours, and that have in common the persistent, compulsive and uncontrolled use of a drug or an activity" (Goldman et al. 2005: 521). "Addictions are moderately to highly heritable" (Bevilacqua and Goldman 2009: 359). According to the National Institute on Drug Abuse, the abuse of tobacco, alcohol and illicit drugs costs the United States more than $740 billion annually in costs related to crime, lost work productivity and healthcare.[3] A virtuous polity would certainly be motivated, by the virtues of benevolence and justice, to redress the problems of addiction that commonly plague contemporary societies. And the exercise of the "intellectual virtues" would encourage us to pay attention to the role played by both genes and environment in our susceptibility to different types of addiction.

Genetics clearly play an important role in our susceptibility to addiction. And a better understanding of this, and ways to perhaps intervene in our biology to reduce that susceptibility, is very important. But it is equally important to recognize how significant is the role of the environment in the risk of addiction, and many environmental factors are currently within our ability to influence (if not control).

Humans can be addicted to a diverse array of things: tobacco, alcohol, prescription drugs such as painkillers, cocaine, and also gambling, sex, work and the internet. There is an innate link to environmental factors because addictions can only develop when there is *repeated exposure* to an addictive substance. If particular drugs are more prevalent in urban settings, for example, we can expect addictions to be more widespread in urban communities than in rural ones. Where a person lives, not just the genes they inherit, matters.

A person's family life also matters. Caring and attentive parents or siblings can help prevent drug use or, better, support intervention and recovery. The stress to which a person is subjected is also an important environmental factor. The stress of job loss, chronic pain, divorce or the death of a loved one can lead people to start abusing alcohol or painkillers. And, finally, peer pressure can be a major factor, especially among teens, in starting to smoke and eventually developing an addiction to nicotine. Peer pressure can also influence the prospects of overcoming an addiction. If a recovering drug addict's spouse or friends are drug addicts, this will reduce the prospects of the individual being able to remain clean in the long term.

A virtuous polity will pursue a wide array of policy measures to help prevent and treat addiction that have nothing to do with altering people's genes. Some addictive substances should be made illegal (cocaine) or regulated (alcohol and tobacco). A legal minimum age requirement for purchasing alcohol and tobacco can help reduce the risks of adolescents becoming addicted. Furthermore, governments can tax alcohol and cigarettes, as the costs of these products will also influence how often they are used (and abused). Some addictive substances, such as strong painkillers like percocet, can only be obtained by a prescription from a doctor. And, finally, a virtuous polity will pursue educative initiatives to raise public awareness about the risks of addictive substances, as well as offering support to addicts and their families.

While heredity is an important risk factor for addiction, environment is also influential and something that is malleable to public policy. As such a virtuous polity will pursue a host of diverse initiatives and policies to try to prevent and treat addiction.

IV Future generations and the non-identity problem

Insights from epigenetics raise not only concerns of *intra*generational justice (e.g. how we can prevent harm to, or promote more equality and fairness for, those currently living) but, as the example of the impact of the Holocaust made vivid, also concerns of *inter*generational justice. A polity could pursue many actions today that are unvirtuous because they fail to take seriously the interests of those who will come into existence in the future. The following two examples reveal how intellectual and moral vice can arise when the current generation acts irresponsibly by failing to consider the interests of future generations.

> *The case of food shortage*: Imagine that polity X, at time T1, decides to aggressively pursue a policy of rapid urbanization because it believed this would bring quick financial benefits. After a few decades of pursuing this initiative, the new generation of citizens living in X now face severe food shortages, even starvation. Large parts of once fertile agricultural land had been quickly converted into profitable high-rise apartment buildings, and the productive workers needed to provide the labour necessary to achieve a sustainable food supply for the population had dwindled significantly. The sharp rise in the urban population, coupled with irresponsible food production public policies, meant that the polity could no longer meet the basic material needs of its population.

In such a scenario, the decision to implement policy X, at time T1, would be unvirtuous. Policy analysts should have run projections on the impact of food production and pursued economic growth and development in a manner that did not erode the demands of benevolence and justice.

Consider another example where concerns of intergenerational justice can arise – climate change.

> *Climate change*: At time T1 our polity has reason to believe that our continued reliance on fossil fuels will lead to further increases in global warming. And a warmer climate will most likely lead to more flooding, higher risks of infectious diseases such as malaria, and more frequent and intense natural disasters such as hurricanes. But our polity continues to rely on fossil fuels, refusing both to adapt to be prepared for global warming *and* to reduce carbon emissions to help prevent the worst effects of climate change.

The example of climate change is a stark reminder of how important it is for us to consider the impact actions taken in the "here and now" will have on those who will live in the future. Advances in our understanding of our biology, and the prospect of purposively intervening in that biology through genetic or environmental interventions, raise a host of complex considerations.

Researchers have long known that an expectant mother's exposure to certain environmental factors, such as smoking, diet and the use of elicit narcotics, can affect fetal development (and thus, if carried to term, that of the child). In order to gain a better understanding of transgenerational epigenetic effects (such as DNA methylation) on future generations, by ruling out environmental influences during gestation, scientists have recently turned their attention to fathers and the epigenetic effects of sperm.

The life history of a parent (e.g. their nutrition and exposure to toxins) can influence, in both negative and positive ways, the development of their offspring. And this means that the moral stakes at risk in ensuring citizens can enjoy the environments conducive to their developing as healthy and happy people extend beyond the confines of just concerns of *intra*generational justice. They also raise concerns of *inter*generational justice, given that the life history of the current generation, and even past generations, influences the biology of future generations.

For example, research suggests that the food available to men during their pre-pubertal growth phase is associated with the risk of diabetes and cardiovascular disease and mortality in their grandsons (Kaati et al. 2002). And, at least in mice, paternal exposure to cocaine may cause impairment of the working memory of their offspring (He et al. 2006). And, finally, the age of a man also impacts the IQ of his offspring. One study of humans found that parental age accounted for approximately 2 percent of the total variance in IQ scores, with later paternal age lowering non-verbal more than verbal IQ scores (Malaspina et al. 2005). The age of the father also influences the probability of their offspring having autistic-like traits (Lundstrom et al. 2010).

These findings suggest that there is "epigenetic inheritance" – that is, that offspring can inherit altered traits due to their parents' (and more distant ancestors') age and past experiences. Such findings expand the traditional purview of "procreative ethics," which tended to focus solely on those cases where an expectant mother's behaviour could seriously affect fetal development. To illustrate how this is the case, consider the following examples addressed by Buchanan and his co-authors in *From Chance to Choice*, for doing so offers some helpful instruction on how we might tackle the moral concerns raised by "epigenetic inheritance."

The examples addressed concern maternal procreative and parental decision-making, but they are instructive for how we might assess the cases of paternal epigenetic inheritance. Here are the three scenarios Buchanan et al. (2000: 244) invite us to consider.

P1: A woman is told by her physician that she should not attempt to become pregnant now because she has a condition that is highly likely to result in moderate mental retardation in her child. This condition could be prevented if she delays becoming pregnant for a month and takes medication. But she refuses to take the medication, gets pregnant now and has a child who is moderately retarded.

P2: A pregnant woman is told by her physician that she needs to take some medication in order to prevent her baby being born mentally retarded. She fails to take the medication and the baby is born moderately retarded.

P3: A three-year-old child is in need of immediate medical treatment. If the child does not receive this treatment she will become mentally retarded. The mother prevents the child from undergoing this treatment, thus resulting in her child becoming moderately retarded.

Do we feel the prospective mother, and in the case of P3 the mother, has done something morally wrong in all three cases? For many people their intuitive response is, "yes," the mother has committed a wrong in all three cases. But because the timing of the intervention is different in all three instances, the basis for claiming that a moral wrong has been committed will be different.

P3 is the easiest case where the basis of the moral wrong can be articulated because there is clearly an identifiable *person* who is harmed by the mother's inaction – namely, her child. Had the mother taken the child to get the necessary medical treatment, the child would not have suffered mental retardation. Parents are *morally required* to care for their children, which entails taking reasonable precautions to help prevent them from suffering harms such as starvation, neglect and, in this case, moderate retardation. This moral duty to prevent harm is so stringent that it can be legally enforceable. Parents can lose custody of their children if they consistently fail to provide them with a safe living environment.

P2 is arguably a more tricky example to address because the issue of the moral status of a fetus is extremely contentious. Some may argue, perhaps on theological grounds, that a fetus should have the same moral status as a person, whereas others might argue that a fetus (at least one in an early stage) does not have any moral status. This metaphysical debate can be side-stepped by noting that, assuming the mother is intent on carrying the fetus to term, her actions adversely impact, albeit prenatally, the child that eventually comes into existence. That child could have existed without moderate retardation had the mother taken the prescribed medication.

In both P3 and P2 the moral wrongfulness of the mother's action are captured by what Larry Temkin (1987) calls the *person-affecting principle*, which maintains:

1 one situation is worse (or better) than another if there is someone for whom it is worse (or better), and no one for whom it is better (or worse), but not vice versa, and
2 one situation cannot be worse (or better) than another if there is no one for whom it is worse (or better). (Temkin 1987: 166–7)

In both P3 and P2 the situation is worse for the child and the eventual child, respectively. Because in P3 the mother prevents the child from receiving medical attention, the child is worse off when they develop moderate retardation compared to how they would have fared had they received the medical intervention. And in P2 the child is born with mental retardation that could have been avoided had the mother taken the medication during the fetal development stage. Because, in the case described by Buchanan and his colleagues, the mother's reason for not taking the medication was not based on some morally defensible logic (e.g. to avoid inflicting some adverse risk on her own health), it is a situation in which one is worse off (the child born) and no one is better off than the alternative scenario of the mother taking the medication.

But the person-affecting principle cannot explain what is morally wrong in P1 because it is committed to premise (2) – one situation cannot be worse (or better) than another if there is no one for whom it is worse (or better). What is distinct about the case of P1 is that it is a scenario where the identity of the child born is influenced by the decision whether to take the prescribed medication or not. If the mother waits a month before trying to conceive and takes the medication, as advised by her doctor, she would conceive a different child than the potential child she could conceive now. This scenario raises the problem known as *the non-identity problem*, an issue that has received considerable attention from philosophers (Parfit 1984; Hanser 1990; Heyd 1992; Roberts and Wasserman 2009; Boonin 2014). The non-identity problem arises in cases where the people born as a result of the actions or policies being morally scrutinized would not have been born at all had the alternative action been taken. In such a case, assuming the people born have lives worth living, we cannot describe the action or policy pursued as being wrong because it harmed the people born from such actions, as the only alternative was for them not to exist.

P1 is a challenging scenario to address because the non-identity problem arises. Most people have the intuition that some wrong has been committed. But *who* has been wronged? The child born with moderate retardation cannot be said to be wronged because if the mother did not conceive when she did she would have had a different child. Assuming the condition with which the child is born is not worse than "non-existence," it seems that, at least according to the person-affecting principle, there is no moral wrongdoing in P1.

Buchanan et al. suggest that one way of articulating our moral convictions that wrong takes place in P1 is to endorse a *non-person-affecting principle* that applies, not to distinct individuals, but to classes of persons who will exist if the suffering is, or is not, prevented. The authors endorse the following non-person-affecting principle:

> Individuals are morally required not to let any child or other dependent person for whose welfare they are responsible experience serious suffering or limited opportunity or serious loss of happiness or good, if they can act so that, without affecting the number of persons who will exist and without imposing substantial burdens or costs or loss of benefits on themselves or others, no child or other dependent person for whose welfare they are responsible will experience serious suffering or limited opportunity or serious loss of happiness or good. (Buchanan et al. 2000: 249)

New findings about the influence of epigenetic inheritance on future generations raise interesting questions about the role of benevolence in our procreative decisions. Virtuous parents would obviously seek to prevent harm to the specific individuals in the P2 and P3 examples. Such examples are "same-person choices." Because P1 is not a "same-person choice" – that is, if the mother decides to delay conceiving and takes the medication she will give birth to a different child – our moral disapproval of her actions in P1 must stem from a commitment to the belief that *the class of potential persons* that constitute her "offspring" warrants some moral consideration in her decision-making. And in P1 she fails to demonstrate appropriate moral concern.

Moving from a moral examination of the mother's actions in the three examples posited by Buchanan et al., let us consider the paternal examples of epigenetic inheritance mentioned earlier. Some instances of epigenetic inheritance will fall outside the realm of something for which a potential parent could be responsible – for example, the case of the food available to men during their pre-pubertal growth phase that affected their grandsons. But other cases clearly do fall within the realm of actions for which we can hold persons morally responsible, such as using cocaine and thus risking the impairment of the working memory of one's future offspring. And the case of paternal age is a tricky variable, as benevolence (or the person- and non-person-affecting principles) does not place *undue* burdens on prospective parents to benefit their potential offspring. Potential parents should not be expected to align their life and procreative aspirations around the duty to confer the greatest health benefits to their offspring.

Curley et al. (2011) explain how there is an inverted U-shaped relationship between parental age and offspring health:

> Increasing age of fathers has been found to be related to elevated rates of schizophrenia (Malaspina et. al., 2001), autism (Lundstrom et al., 2010; Reichenberg et al., 2006), and early-onset bipolar disorder (Frans et al., 2008) in offspring, as well as with reduced IQ (Malaspina et al., 2005) and social functioning in adolescents (Weiser et al., 2008). Furthermore, very young paternal age (typically under 25) is also associated with negative outcomes such as decreased IQ (Auroux et al., 2009; Auroux et al., 1989; Malaspina et al., 2005), higher rates of autism-spectrum disorder (Lundstrom et al., 2010) and an increased risk of birth defects (Mcintosh et al., 1995), suggesting that there exists an inverted Ushaped relationship between paternal age and offspring health. (Curley et al. 2011: 308)

Does the moral concern to promote the health of our potential offspring mean that prospective fathers (or mothers) should try conceiving only during those ages where the effects of epigenetic inheritance are likely to confer the *best* health prospects on their potential offspring? I would argue that the answer is clearly "no." There are many other considerations that come into play when people make decisions to become parents. A prospective parent might wish to wait until they find the right long-term partner – someone whom they believe will make a good partner for themselves and co-parent for their children. Or they might wish to wait until they are more financially stable or have progressed far enough in their career advancement, so that they have the resources and time to invest more in parenting. These decisions can also positively impact their offspring in important ways. Prospective parents should not be expected to undertake *unreasonable burdens* to confer epigenetic benefits to their potential offspring – such as rushing to select a mate or forfeiting possible career advancement to have children earlier. Of course these are pressures faced by every potential parent currently, as age does decrease fertility. But I do not think that the findings of epigenetic studies should dramatically increase the moral demands on prospective parents.

If a new intervention were developed to be taken by older prospective fathers that was safe and not very burdensome – say analogous to prospective mothers taking folic acid – and which could reduce the elevated rates of schizophrenia, autism and early onset bipolar disorder in their offspring, then there would be strong grounds for claiming that utilizing such an intervention is *morally obligatory* to prevent unnecessary harm to their offspring. All sexually active men below and above the age of the most optimal epigenetic inheritance could be encouraged to take advantage of the intervention. This could confer significant health benefits on future generations. The moral stakes involved would be very different if we move from a situation of simply gaining more knowledge about the role of epigenetic inheritance to being able to manipulate such inheritance in a safe and non-invasive manner.

Insights from epigenetics reveal the folly of narrowly committing a polity to the goal of trying to alter DNA sequences only (rather than the expression of genes) to improve the opportunities for individuals and a population to flourish. Such a strategy presumes *genetic determinism*, which is a faulty understanding of our biology. The intellectual virtues instruct us to pay attention to the relevant facts. And what we are learning from epigenetics is that gene expression can be influenced by environment. It is true that some health disadvantages are caused solely by specific genetic mutations, and thus the appropriate way of remedying those disadvantages might be to alter specific genes, be it via gene therapy or gene editing. But environment is also extremely important and can offer a way to help prevent or treat disadvantage. As is illustrated by the cases of poverty, the Holocaust, addiction, and the epigenetic effects of sperm, environmental exposures – physical, chemical and emotional – also profoundly influence the life prospects of the current generation, as well as those of future generations.

Rather than distracting a virtuous polity away from the traditional concerns of distributive justice and public health, a better understanding of how environment and genes interact to influence our health and wellbeing should help us better realize the demands of benevolence and justice. Theories of distributive justice have tended to focus either solely on what John Rawls (1971) called the "social primary goods" – things such as wealth and income – *or* on access to healthcare provisions (e.g. access to a doctor, medicine, etc.). An integrative approach is what is needed. Advances in our understanding of genetics reveal the importance of taking an expansive and informed perspective on *all* the factors – natural and social – that influence our health and wellbeing. Prioritizing one particular distributive principle or goal (such as equal wealth for all or a genetic decent minimum) is likely to skew this expansive perspective and oversimplify the complex nuances of the biological processes that shape and influence our life prospects.

Discussion questions

1 Some may worry that the attention given to new findings about the important role genes play in addiction, anxiety, etc., will draw attention away from helpful environmental interventions (e.g. poverty reduction, cognitive behavioural therapy, etc.) as the prospect of genome editing could be presented as a potential "cure-all" solution to such problems. Do findings in epigenetics help alleviate such worries because they reveal how important environment can be in gene expression? What findings do you think are the most significant insights and why?

2 What do you think of the *nature vs. nurture* debate? In what ways has it been a helpful discussion? And in what ways has it been a hindrance?

3 Concern for future generations raises a host of complex moral considerations, ranging from the stringency of such duties to the non-identity problem. Do you think scientific findings from epigenetics, for example those concerning paternal epigenetic inheritance, alter the moral landscape of "procreative ethics" in an interesting and important fashion? If so, how do they impact this landscape?

5

Reproductive Freedom

I Introduction

The bioethicist Dan Brock (2005) provides a helpful, and thorough, moral analysis of three distinct grounds for granting parents expansive reproductive and parental freedom – *autonomy*, *wellbeing* and *equality*. In this chapter I develop a virtue-oriented analysis of reproductive and parental freedom by expanding upon these three grounds, with the goal of elucidating how a virtuous polity might approach the regulation of technologies such as pre-implantation genetic diagnosis (PGD). For example, would a virtuous polity permit parents to make use of new genetic tests to screen embryos for medical (e.g. reducing the likelihood of a genetic disease) and non-medical purposes (e.g. sex selection of embryos for family balancing)? While virtue ethicists can place significant weight on all three of the values of autonomy, wellbeing and equality, what is distinctive about a virtue-ethics approach, rather than a principle-oriented one, is that none of these three values will be treated as "sacred" or "inviolable" or ranked as higher than the others. Virtue requires taking a *purposeful*, and *provisional*, stance on the weight to attribute to different values and searching for fair accommodations of such values when, as inevitably happens in the non-ideal world, conflicts among them arise.

Autonomy is the value of letting people be authors of their own lives. The value is given a primacy in the history of liberal political theory by such thinkers as John Stuart Mill. Mill wanted to constrain the extent to which government, and society more generally, might interfere with the actions of persons, who should be treated as free and equal. In *On Liberty*, Mill famously proclaimed that "The only purpose for which power can be rightfully exercised over any member of a civilized community, against his will, is to prevent harm to others. His own good, either physical or moral, is not a sufficient warrant" ([1859] 1956: 23). Permitting citizens to live an autonomous life means letting them risk making bad decisions. Mill had faith in human rationality that, over time at least, most people, when they have the opportunity to exercise autonomy, will come to learn to make better decisions concerning how to live their lives compared to what might happen in a paternalistic society, where the government attempts to make personal decisions for them.

Autonomy, on the moral schema deployed in this book, falls under the purview of the moral virtue of justice. The just society will place significant (though not absolute) weight on the importance of permitting citizens to live autonomous lives. This treats persons as equals and helps them flourish as individuals, as well as enabling a polity to flourish as a collectivity.

When it comes to issues pertaining to procreation, a virtuous polity will grant citizens an expansive range of autonomy. Individuals can decide for themselves (1) if they even want to try to become a parent; (2) with whom they procreate; (3) when to have children; (4) how many children to have; and finally, once a child is born, (5) how to raise their child.

Because (5) involves directly, indeed profoundly, influencing the life prospects of others (namely, one's offspring), it is subject to some clear limitations. Parents cannot neglect or abuse their children. A virtuous polity would not justify parental neglect by invoking the defence "respect parental freedom!" The moral virtue of justice requires that those who are vulnerable, especially children, be protected and cared for. In most cases children's parents are the persons best suited and motivated to nurture and care for them. Thus the values of respecting parental freedom and protecting the interests of children will often align. And in those unfortunate circumstances where that alignment does not occur and parents are neglectful or abusive, the government ought to intervene to protect the interests of children – for example, by providing the parents with additional support (e.g. counselling) or, if necessary, placing the children in foster care.

II Parental wellbeing

The second value that Brock identifies as the moral basis of the rights to reproductive freedom is wellbeing. Having control over our own decisions to have children, or how many to have and with whom, has a profound impact on our wellbeing. Governments do not have to mandate "fertility targets" for couples because most humans desire to become parents. According to a recent Gallup poll in the United States, "more than nine in 10 adults say they already have children, are planning to have children, or wish that they had had children."[1] Legislating that people must have children, or that adults cannot have more than a certain number of children, or that they must abort a fetus deemed "unworthy" by an intrusive eugenic policy would threaten the wellbeing of citizens and be grossly unjust.

Brock claims that "it is plausible to suppose that when persons wish to have and raise children, doing so generally contributes in very substantial ways to their well-being" (2005: 383). But this point is not as straightforward as Brock presumes. If wellbeing is equated with "happiness" rather than just the stated preference noted in the Gallup poll, the story of whether having children promotes a parent's happiness is an interesting one but also complex and contentious.

"Happiness" is itself a contested concept. Researchers who study, and measure, happiness might deploy "experiential sampling" to get reports of how happy parents are when performing certain tasks – such as feeding their kids, cleaning the house, etc. Because women typically have, and continue to do, the lion's share of such duties it is not surprising to see that fathers report higher levels of satisfaction than mothers (Nelson et al. 2013).

A different way to measure happiness is to employ "retrospective evaluation" where people are asked to reflect upon such achievements as parenting their offspring through childhood and adolescence. I have three sons myself, and when I reflect on the joy and happiness they have contributed to my life it is second to none. Admittedly fathering is not all "joy and happiness." A sizable portion of my income goes to clothing, feeding and supporting my kids. And I have spent countless hours at their sporting events, sleepless nights tending to them when they were sick, helping them with homework, making their lunches, doing their laundry, etc. But these costs are, at least for me, well worth that price because they have taught me so much about myself and life, and I am grateful for having them in my life. They have made me more patient and understanding. They remind me of the importance of "being in the moment," I learn new things about our culture (e.g. the latest "memes" or new music) from them, I meet new people because of them, etc.

Living a meaningful life involves attachment to something greater than oneself. Being a father provides me with an opportunity to be attached to something larger – namely, my family – and this provides me with a deep and lasting sense of purpose and meaning. I also find profound fulfillment in my teaching and research and in volunteer work with children and inmates. But my reproductive freedom has been an important liberty for me personally, one that has permitted me to find fulfillment in the unique joys of parenting.

Of course my experience of the joys of parenting can be different from other people's. I have a secure and well-paid job, and my children are very well behaved (most of the time – I won't mention the teen years!). These factors can have a significant impact on one's fulfillment from parenthood. In "Parenthood and life satisfaction: why don't children make people happy?" Matthias Pollmann-Schult (2014) comes to three conclusions:

1 parenthood by itself has substantial and enduring positive effects on life satisfaction;
2 these positive effects are offset by financial and time costs of parenthood; and
3 the impact of these costs varies considerably with family factors, such as the age and number of children, marital status, and the parents' employment arrangements.

The importance of the wellbeing of parents is very relevant when considering the ethics of utilizing pre-implantation genetic diagnosis (PGD). This procedure involves screening embryos, after undergoing IVF, to filter them for specific diseases, for example, before deciding on which ones to implant. Consider the use of PGD for medical purposes. Suppose a couple is at a high risk of a serious single-gene disorder. Perhaps they have already had a child who died early in life from a genetic disorder, or they have undertaken genetic testing and know they are carriers for a condition that could be passed on to their offspring. The couple decide to conceive this time via IVF and would like to screen embryos for the risk of genetic disease, choosing to implant embryos that do not indicate the presence of the genes for specific single-gene diseases.

Would a virtuous society permit parents to utilize such screening measures? Benevolence prescribes that we should aspire to benefit others. In this case, the benefits in question are the potential benefits to the parents of having healthy children and avoiding the extra caregiving burdens of having a child with a genetic disease. Caregivers of children with chronic illness, for example, reported significantly greater general parenting stress than caregivers of healthy children (Cousino and Hazen 2013). Are such benefits weighty moral interests? I think the answer is clearly "yes" for the most serious genetic disorders. The parents will be the ones expected to provide the extra care and attention to a child with serious medical needs. If the parents aspire to reduce the likelihood of procreation imposing such burdens on themselves and their family, why would a virtuous polity seek to prevent them from doing so?

When some form of PGD could be reasonably inferred to benefit the wellbeing of prospective parents, there is a presumption in favour of granting the parents the liberty to utilize such technologies unless there is a pressing and substantial reason for constraining such interventions. Utilizing PGD for medical purposes, while somewhat controversial, is something I believe a virtuous polity would address in a manner that is deferential to the autonomy and wellbeing of prospective parents.

What about utilizing PGD to screen for risk of late onset genetic disorders such as hereditary cancer? In such cases the appeal to the interests in reducing the extra caregiving burden on parents does not apply, or at least not to the same degree as with early onset genetic disorders. But prospective parents who might themselves have survived breast or bowel cancer, or had a parent die from such conditions, might be anxious about the risk of passing these conditions on to their offspring. Would a virtuous polity permit such prospective parents to utilize PGD to test for inherited, lower penetrance conditions? Because IVF is an invasive procedure for the woman and is not guaranteed to yield a successful result, it can also add psychological stress. A virtuous polity would have to consider such factors, and ensure that prospective parents are aware of the various risks involved. At the same time it must also be acknowledged that trying to conceive naturally can add psychological stress and that a normal pregnancy has its own risks. A virtuous polity would consider the relevant risks and harms IVF and PGD pose to prospective parents, keeping in mind there are risks and harms with a normal pregnancy. I think there is a strong case to be made for thinking a virtuous polity would permit PGD for late onset, as well as early onset, genetic disorders.

One line of possible objection to permitting PGD screening for genetic disorders comes from disabilities rights advocates.[2] A number of distinct concerns could be raised, but perhaps the most pressing one is what Buchanan and his colleagues call the *Expressivist Objection*. This objection maintains that the development of genetic interventions to prevent disabilities "express negative judgements about people with disabilities, and that these judgements constitute a profound injustice to those people" (2000: 272). And this critique is often conjoined with what the authors call the *Loss of Support Argument*: "as the application of genetic science reduces the number of persons suffering from disabilities, public support for those who have disabilities will dwindle" (ibid.: 266).[3] The Expressivist Objection and the Loss of Support Argument could be utilized to oppose permitting or encouraging screening for disease or disability by arguing that the wellbeing of existing persons living with those conditions is diminished.

One could certainly conceive of a scenario where genetic screening could be utilized in a fashion that would both express negative judgements about people with disabilities *and* diminish support for them to a degree that it would constitute an injustice. For example, suppose a polity decided that the best way to address the health challenges posed by cystic fibrosis (CF) was to offer PGD to all prospective parents who are CF carriers, so they could undergo IVF and implant embryos that did not have the CF gene. But the polity decided that that was the *only* support the polity would offer to address CF, to ensure greater compliance with the screening. It even went so far as to eliminate support for those currently living with CF, and it abandoned all medical research into gene therapy for CF, stating instead that its goal was to prevent people from being born with CF rather than offer treatments or develop a possible cure for the condition.

In such a harsh and unreasonable example, the disabilities rights advocates would have compelling grounds for claiming such a society is unjust – it epitomizes both moral and epistemic *vice* rather than virtue. A virtuous polity would pursue the goal of preventing CF in conjunction with offering support for those currently living with the disorder and those who will still be born with CF, given that offering IVF and PGD will not prevent all CF births as many potential carriers will still procreate naturally (especially unplanned births). In such a far-fetched scenario the concerns about the "loss of support" would clearly be valid. But a virtuous polity would ensure that any screening interventions offered would not erode the demands of justice to offer support to those currently living with CF or any other genetic disorder or disability.

We do not need to construct the fictional case of screening for CF to make this point. As discussed in chapter 1, the Centers for Disease Control recommend that women of childbearing age in the United States take 0.4mg of folic acid every day to help prevent major birth defects of the brain and spine, such as spina bifida. Does this recommendation lead to the loss of support for those living with neural tube defects, and is it an affront to their equality? If the recommendation that woman of childbearing age should take folic acid to help prevent neural tube defects was the *only* response the United States took to such birth defects, it would clearly be unjust and lead to a loss of support. Even if there was universal compliance with taking folic acid (which is highly unlikely), this would not prevent all neural tube defects. So focusing only on the prevention of disability is insufficient. Benevolence and justice also require providing support to those born with, and those currently living with, neural tube defects. Children with spina bifida will require not only medical care and surgical treatments but the provision of braces, crutches or wheelchairs to improve mobility. And they may need extra help from professionals to adapt to life in the school environment, etc. Prescribing folic acid to prevent birth defects is not unjust when such a measure is conjoined with treatment and support to those living with disability.

As for the claim that permitting, or encouraging, prospective parents to screen embryos (or take folic acid) to reduce disability expresses negative judgements about people with disabilities, I think Buchanan and his co-authors formulate a nice response to this concern by drawing a distinction between devaluing disabilities without devaluing people with disabilities. A virtuous polity pursues many measures to prevent disability – requiring people to wear seat belts when in automobiles, to wear safety glasses when working with hazardous materials, to wear helmets when riding a bike, etc. Such measures are predicated upon the conviction that benevolence requires we pursue cost-effective measures to reduce disability. But disvaluing a disability – such as compromised vision or cognitive impairment – does not mean that people who are visually or cognitive impaired are devalued. A society can legislate the wearing of eye protection in certain workplaces *and* still be committed to the equality of all persons – visually impaired or not. The latter is determined not by the extent to which a society aspires to prevent a disability or disease but by the support it offers to those living with the disability.

III Replacement fertility and patriarchy

The third distinct ground for granting parents expansive reproductive and parental freedom that Brock identifies is *equality*. And in this section I wish to expand upon the emphasis a virtue ethics defence of reproductive freedom will place on equality by highlighting once again the importance of the "intellectual" or "epistemic" virtues as well as moral ones. Recall, from the introduction, that these intellectual virtues include the ability to recognize the salient facts and a sensitivity to details, intellectual humility, adaptability of intellect and the detective's virtues (thinking of coherent explanations of the facts).

One important fact of human civilization is that the history of humanity is one of gross violations of reproductive freedom, especially of women. Recognizing both the fact and the history of *patriarchy* is imperative for determining how a virtuous polity will address the issue of the scope and limits of reproductive freedom. Understanding the history of egregious violations of reproductive freedom is important as it will inform a virtuous polity's response to the societal dilemmas that arise when reproductive freedom conflicts with other pressing societal aspirations.

As was noted in chapter 1, "Eugenics: Inherently Immoral?," there is an important historical lesson to be learned by looking at the attitude of past eugenicists who advocated unjust measures to "improve the biological character of a breed." Paul Popenoe, who founded the American Institute of Family Relations in 1929 and championed exclusivist eugenic aspirations, had this to say about eugenics:

> Eugenics rests on two axioms so simple that a child can understand them. If a people is to survive, it must produce in each year, or each generation, enough children to take the places of those who die during that period. And if it is to avoid deterioration which would also prevent survival, it must encourage childbearing from the part of the population that is, in general, fit, rather than predominantly from the mentally diseased, the mentally deficient, and the physically defective. (Popenoe 1935: 451)

The second axiom that Popenoe describes clearly reflects the prejudice, exclusivity and ignorance of his time and culture. While a virtuous polity would not necessarily rule out the aspiration to "improve its biological character" (Russell 1929) – indeed, I argued in chapter 1 that it can view such aims as morally *required* – a virtuous society would denounce as unjust the second eugenic axiom endorsed by Popenoe. The first axiom of eugenics Popenoe mentioned, that reproduction is essential to society's survival, raises the concern of "replacement fertility," and this is a reality worth addressing in further detail as it helps provide some historical context for the importance of Brock's third ground for reproductive freedom – equality.

In chapter 3 we addressed Bruce Carnes's (2007) explanation of the disposable soma theory of aging, and this is something that relates to Popenoe's first axiom of eugenics. Let me briefly repeat Carnes's explanation. The world is a dangerous place. Death is, for humans and all other living things on this planet, inevitable. For a species to survive a solution to this problem must be found. And, for sexually reproducing species, that solution is reproduction. If a species can win the race between reproduction and death, then they will survive until that time when they lose that race. And then they become extinct. Most species that ever lived on this planet are now extinct. The dodo bird, dinosaurs and the Neanderthals all lost the race between death and reproduction.

This reality of life of this planet means that all human societies, from hunter-gatherer communities to modern liberal democracies, must ensure that its members reproduce at a rate that guards against its becoming extinct. Dwindling population size, especially of healthy productive members, will leave a community vulnerable economically and militarily. The evolution of our species and the development of our cultures were profoundly shaped by the reality that reproduction is such an important collective imperative.

Historically, virtue ethicists such as Plato proposed measures like censorship and eugenic selective breeding to ensure the conditions necessary for a virtuous polity to flourish could be realized. But Plato's elitist conception of justice was one that prescribed that each person should do what they are best suited to do. As such, the measures he proposed to ensure people fulfilled the task for which they were suited entailed gross violations of autonomy. The modern virtue of justice championed in this book is one that posits significant weight to the value of permitting persons to have control over their bodies and minds, including what they think and how they express themselves, and especially surrounding issues to do with procreation and parenting.

The freedom to decide if we want to have children, how many to have, how to raise them, etc., is a very recent development in liberal democracies. Historically, such expansive autonomy over procreative decisions was unthinkable (and that is still the case today in many parts of the world). To understand why, we must appreciate the precarious circumstances (e.g. high mortality rates) in which earlier periods of human civilizations had to function.

If a polity were to remain the same population size over time, it would have to achieve what demographers refer to as "replacement fertility." Replacement fertility is the level of fertility at which a population exactly replaces itself from one generation to the next (Craig 1994). Suppose, for example, that there were no early or mid-life mortality in a polity, and this polity was a "closed" society (meaning there was no migration of people in or out of the country). To achieve replacement fertility, an average of two children would replace all mothers and fathers provided the same number of boys and girls were born and all of the female offspring survived to the end of reproductive age. Of course no polity in the real world could eliminate childhood or adolescent death. This means a fertility rate of below two will be less than replacement level. The common number used by demographers to define replacement fertility level in developed countries is 2.1 children per woman (ibid.). In today's world, the reality of population aging and declining fertility means that polities face significant challenges in terms of the impact this demographic transition will have on the economy, healthcare and pensions.

Historically, for most of our evolutionary history, replacement fertility levels had to be much greater than 2.1 because of very high mortality rates. This reality shaped the biology of males and females, which in turn shaped human culture. The importance this reality has for women is highlighted in the feminist historian Gerda Lerner's masterful analysis of the history of patriarchy in *The Creation of Patriarchy*. Lerner notes that, "because of the extreme and dangerous conditions under which primitive humans lived, the survival into adulthood of at least two children for each coupling pair necessitated many pregnancies for each woman" (Lerner 1986: 41). This reality sets the stage for the evolution of patriarchy.

Patriarchy is the "manifestation and institutionalization of male dominance over women and children in the family and the extension of male dominance over women in society in general" (Lerner 1986: 239). Lerner argues that patriarchy is first created out of *necessity*. It begins with the establishment of an initial division of labour by which women do the mothering (i.e. nurturing and raising of offspring) – a division of labour that was necessary for group survival for millennia (ibid.: 40).

Biological differences between males and females reflect this "sexual division of labour." Females must invest a lot more time and energy, relative to males, in reproduction in order to pass on their genes. After she becomes pregnant, an expectant mother's body undergoes drastic changes, as it diverts food and nutrients to her potential offspring. But, for men, the needs and strategies for managing energy are different. Men do not menstruate, breast feed, etc. Investing more energy in body size and muscle development can increase reproductive success for males given the competition they face (e.g. fending off rival competitors and also attracting mates). As Richard Bribiescas argues in *Men: Evolutionary and Life History*, "unlike mothers, fathers are not *required by their biology* to provide child support. Every calorie ingested by a human male is his to keep – and to invest, if he sees fit, in pursuits other than protecting and provisioning the younger generation" (Bribiescas 2006: 221).

Unlike female fertility, which is limited by the amount of energy involved in having offspring, male fertility is limited by the number of available mates. The "male warrior hypothesis" (McDonald et al.

2012) argues that, for men, intergroup conflict represents an opportunity to gain access to mates, territory and increased status, and this may have created selection pressures for psychological mechanisms to initiate and display acts of intergroup aggression. In addition to the imperative to buttress the resources available for fighting to acquire scare resources, early human societies had to invest sufficient resources in the reproductive and caring labour necessary to bear and raise healthy offspring. It was thus "absolutely essential for group survival that most nubile women devote most of their adulthood to pregnancy, child-rearing and nursing" (Lerner 1986: 41).

But this initial sexual division of labour, which was (historically) necessary and beneficial to both men and women living in early hunter-gatherer societies, changed in just the past few thousand years. Lerner's study focuses on Mesopotamia from approximately 3100 BC to 600 BC.

> Women's sexual subordination was institutionalized in the earliest law codes and enforced by the full power of the state. Women's cooperation in the system was secured by various means: force, economic dependency on the male head of the family, class privileges bestowed upon conforming and dependent women of the upper classes, and the artificially created division of women into respectable and not-respectable women. (Lerner 1986: 9)

Male dominance over women, according to Lerner, is thus not "natural" or biological; rather, it is the product of a particular historical period in our history.

Adopting an evolutionary analysis of the creation of patriarchy, Barbara Smuts contends that the origin of patriarchy goes back to our pre-human past and "is a product of reproductive strategies typically shown by male (and, to a lesser extent, female) primates" (1995: 20). Smuts argues that the following six factors influenced the evolution of human gender inequality:

1 a reduction in female allies;
2 elaboration of male–male alliances;
3 increased male control over resources;
4 increased hierarchy formation among men;
5 female strategies that reinforce male control over females;
6 the evolution of language and its power to create ideology. (Ibid.)

This brief outline on the arguments of Lerner and Smuts is necessary for developing an ethical analysis of reproductive freedom in the twenty-first century because they provide a foundational starting point for a virtue-oriented analysis of the scope and limitations of reproductive freedom – namely, that the reproductive strategies employed (by men) in the past have been ones that controlled and oppressed women. The moral virtue of justice requires we aspire to treat all persons, regardless of their sex, as *equals*, and this entails redressing (rather than exacerbating) the injustices of patriarchy.

One fundamental goal of a virtuous polity is to determine how it can meet sustainable fertility rates in a way that does not perpetuate patriarchy and/or limit the freedom of women. Let me provide one clear, though no doubt contentious, example of *vice* to illustrate this point. Abortion is a highly contentious ethical issue, and I will not address the question of the moral status of the fetus as that will take us too far afield. But imagine a polity that, faced with below replacement fertility levels and little immigration, decides to prohibit all abortions as a way of boosting fertility rates. Any law prohibiting abortions obviously has a much more profound impact on the freedom and wellbeing of women than men by the very fact that it is women who would have to carry to term an unwanted pregnancy and endure the painful and risky procedure of childbirth.

A virtuous polity would recognize that, "in order for any procreative decision to have ethical significance, the woman involved must have moral agency, authority, and freedom" (Overall 2012: 10). A prohibition on abortion to increase fertility rates is an example of *moral vice* rather than virtue because it compromises the moral agency, authority and freedom of women. A virtuous polity would aspire to meet the challenges of a diminishing population size by encouraging both immigration and higher fertility rates by pursuing effective policies that actually promote (rather than contravene) the equality of women – measures such as providing financial incentives to families (e.g. tax relief), affordable childcare, parental leave, and quality public education for children and adolescents. Such measures aim to serve multiple laudable purposes at once – helping achieve sustainable fertility rates, but also improving equality of opportunity, fostering familial values and education, etc.

IV Procreative beneficence: why not the best?

Julian Savulescu is an influential bioethicist who champions what he calls the principle of procreative beneficence (hereafter referred to as PB). PB states that "Couples (or single reproducers) should select the child, of the possible children they could have, who is expected to have the best life, or at least as good a life as the others, based on the relevant, available information" (Savulescu 2001: 415). It is important to note that PB does not prescribe that prospective parents should consider having an abortion if prenatal tests indicate that a fetus has a genetic disease or disability. Rather, the principle applies to screening embryos at the pre-implantation stage of IVF. PB instructs prospective parents to use all the available information (not limited to just screening against disease) and choose the embryo that is expected to have the best life. For example, when utilizing IVF, more eggs may be fertilized than will be implanted in a woman's uterus. PGD (pre-implantation genetic diagnosis) can be utilized to screen those embryos for information about the likelihood of disease and non-disease traits developing in the potential offspring. When selecting which of the embryos to implant, PB prescribes parents should utilize all available information on the potential for health and other characteristics relevant to the future wellbeing of their potential offspring.

For Savulescu, prospective parents *should* (this is a moral, not legal requirement) make use of all the relevant information available to make informed procreative choices. And the "should," he explains, follows from the fact that they "have good reason to" (Savulescu 2001: 415). Elaborating further on the implications of PB, Savulescu and Kahane (2009: 276) assert that there "*is* reason to obtain and use all genetic and other information about disease susceptibility *and* non-disease states to make a decision to select the most advantaged child."

Perhaps the most pressing objection to PB is the fact that it ignores the concerns of equality that need to be considered when addressing procreative ethics. As the feminist bioethicist Christine Overall notes, "any discussion of the ethics of procreation must include feminist perspectives because choosing whether to have children is gendered; it cannot be discussed as if men's and women's very different roles are irrelevant to the issue" (2012: 8). By prescribing that prospective parents must select "the best," this imposes an expectation that children will be conceived via IVF rather than naturally (de Melo-Martin 2004: 73). Prospective fathers pursuing IVF need only provide a sperm sample. But prospective mothers will first begin to take hormone treatment to control egg production. Mature eggs are then collected via needle aspiration, a procedure that is typically undertaken with pain medicine and sedation. While this surgery is minor, both it and the anesthesia carry potential risks.[4]

One compelling reason for resisting the demands of PB is the fact that there is an *inequality in the burdens* that are imposed on prospective fathers and mothers in terms of what is expected of them if they make use of IVF. Thus characterizing natural procreation as somehow morally "sub-par" adds further burdens and pressures on women, but not on men. And this is unfair, as women already bear a disproportionate extent of the burdens of reproduction. To insist that a couple should pursue IVF and PGD to ensure they select "the best" is to impose more risks and burdens on women, whereas the man needs only to masturbate to provide the sperm.

IVF and PGD are also costly interventions. The fixation of PB on "selecting the genetic best" neglects other important moral and prudential considerations. For example, if prospective parents are intent on spending thousands of dollars (thousands *more* – parenthood is extremely expensive as it is) to improve the opportunities for their offspring to flourish, perhaps that money would be better spent on *environmental* interventions (extra tutoring, etc.) rather than on screening the genes of embryos. Furthermore, prospective parents will have other obligations and commitments in their lives to pursue beyond just parenting. Do they have to forfeit all their savings or future vacations in order to pursue conceiving via IVF so they can screen for "the best" possible embryos to implant? Are they failing as loving and caring parents if they spend those resources on other things (e.g. themselves or prosocial aims such as charity) and are content to conceive naturally rather than spend thousands of dollars trying to conceive artificially in the hopes that that will create a child with better genes?

The position advocated by Savulescu and Kahane, which could be classified as one arguing that there is a *moral obligation* to genetically engineer our offspring to have the best life possible, stands in stark contrast to the position advanced by Michael Sandel in *The Case against Perfection*. Sandel argues that parents should be accepting of their children and thus not aspire to genetically engineer them.

> To appreciate children as gifts is to accept them as they come, not as objects of our design or products of our will or instruments of our ambition. Parental love is not contingent on the talents and attributes a child happens to have. We choose our friends and spouses at least partly on the basis of qualities we find attractive. But we do not choose our children. Their qualities are unpredictable, and even the most conscientious parents cannot be held wholly responsible for the kind of children they have. (Sandel 2007: 45–6)

So which is the morally right course of action for parents to pursue – utilizing PGD to genetically engineer their children to be "the best" or forfeiting any such innovations? I believe the virtue ethics tradition offers a more nuanced moral lens than that prescribed by PB or Sandel, one that instructs virtuous parents to strive for the mean between the two positions.

Rosalind McDougall (2007), for example, defends a virtue ethics account of parental obligations that champions three parental virtues – acceptingness, committedness and future-agent-focus. I will not pursue a detailed analysis of McDougall's account here, but her emphasis on acceptingness and committedness nicely illustrates how both the *epistemic* virtues of humility and recognizing the salient facts are important to exercising parental virtue, and that virtuous parents will strive for the middle ground between PB and Sandel's construal of "acceptance."

When elaborating on the virtue of acceptingness, McDougall notes that it is an "inherent feature of human reproduction that a child's characteristics will be unpredictable" (2007: 185). The epistemic virtues require parents to understand they cannot control, even if they screen for the "best genes" with PGD, the development of valued phenotypes (e.g. prosocial behaviour, positive emotions, etc.). If taken in isolation, acceptingness might appear to support the same conclusion as Sandel. However, the virtue of committedness entails that a parent is actively engaged in cultivating and nurturing their offspring so they have the opportunity to live flourishing lives. Reading to a child and exposing them to positive influences such as creative thinking, sport and friendships are examples of parental committedness.

One way of understanding the difference between the position of Savulescu and Kahane's defence of PB and Sandel's defence of "acceptingness" is to see the former as an excessive stance on committedness (lacking the sufficient amount of "acceptance"), while the latter is excessively accepting (lacking the sufficient amount of "committedness"). Virtue ethics prescribes a more provisional and nuanced account of the debate. There can be cases when prospective parents are too accepting and cases when they are too committed. To make the elucidation of the virtues of acceptingness and committedness more concrete, consider the following four cases.

> *Case A* Tom and Mary have decided to start a family. They are reasonably affluent and educated and are aware of the fact that, if Mary starts taking folic acid months before they try to conceive and continues during early pregnancy, they could reduce the risks of the occurrence of neural tube defects. However, Tom and Mary decide Mary doesn't need to take folic acid. They will appreciate any child born as a "gift" and do not want to interfere in the natural lottery of life.
>
> *Case B* Tom and Mary have no reason to believe their potential offspring will be at a higher than average risk of genetic disease. Nonetheless, they are motivated by the appeal of PB to pursue IVF and PGD because they want to select "the best" embryos they can rather than leave things to the natural lottery of conceiving naturally.
>
> *Case C* Tom and Mary have already lost a child to an early onset genetic disorder. They decide to undertake IVF and screen out embryos that indicate they have a high risk of genetic disease. Furthermore, if testing for non-disease traits is also available, Tom and Mary would like to select "the best" among those embryos to implant.
>
> *Case D* Tom and Mary have been unsuccessful in trying to conceive a child naturally, so they decide to pursue IVF. Testing for disease and non-disease traits is available, and they decide to use all this information to select "the best" among those embryos to implant.

In Case A, Tom and Mary are overzealous in their commitment to "acceptingness." They could help reduce the probability of their baby being born with neural tube defects by a cheap and non-invasive measure – Mary taking a daily pill in the few months prior to getting pregnant and during the pregnancy. They can easily afford folic acid and they are aware of the fact that folic acid is an empirically sound intervention to reduce birth defects. And yet they do not follow this medical advice because they don't want to treat their children as "objects of their design or instruments of their ambition." And yet the parents' inaction in this case could mean that a child is born with serious medical problems and mobility limitations that could have been avoided had they displayed the right amount of parental "committedness." Virtuous parents would not eschew empirically sound, cheap and non-invasive interventions that could help prevent serious harm to their potential offspring. So I believe Case A is a clear case of parental *vice* rather than virtue.

In contrast to Case A, Case B is an overzealous instance of "committedness." It imposes extra costs (thousands of dollars) and reproductive burdens and risks on Mary to increase the odds of having "the best." "Committedness ceases to be committedness in the parental virtue sense when a parent martyrs himself or herself for the sake of some trivial increase in his or her dependent child's well-being" (McDougall 2007: 187). Why subject a prospective mother to the extra burden of IVF to increase the odds of its yielding a genetically superior child than the child that would result from natural conception? The stakes look different when parents believe there is a higher risk of a genetic disease involved. But, in the case of selecting for non-disease traits by purposively undertaking IVF and PGD to try to influence the natural lottery, it begins to sound as if the parents believe in genetic determinism and have an unhealthy helicopter parent attitude (e.g. "If only we had chosen the best genes for our child, then he or she would have ended up getting into Harvard!"). Humility teaches us that, as parents, we must accept and understand that life involves happenstance and other unknowns that are beyond our control.

Unlike cases A and B, where I think there is an obvious excess of acceptingness and committedness (respectively), cases C and D are different for a number of reasons. In both C and D the extra reproductive burdens and risks undertaken by Mary (by going through IVF) are to avoid the birth of a child with a genetic disease (in Case C) and because of infertility (in Case D). Having already undertaken IVF, the prospective parents have the option of gaining more information about the potential embryos to select and decide to screen for non-disease genes as well in order to select "the best." But doing so does not impose extra burdens on Mary, as in Case B. In Case B the only reason the couple is undertaking IVF is to "select the best," whereas in C and D the reason the couple has pursued IVF in the first

place has nothing to do with selecting "the best." However, once IVF
has been undertaken, using more information to select "the best" is
not making oneself "a martyr" to benefit one's offspring. Unlike in
Case B, the degree of committedness displayed in cases C and D do
not appear to be excessive. Are the decisions made in C and D virtuous? I think this is debatable. But what I think is clear is that cases C
and D are not instances of parental vice (whereas Cases A and B are).

For the virtue ethicist, a moral analysis of the pros and cons of
utilizing PGD to select "beyond disease" will not yield an easy "yes"
or "no" result. Context really matters. Utilizing PGD to select "the
best" is not always morally required (indeed, it can be immoral when
it places unfair burdens on prospective mothers), nor is it always
morally impermissible. Virtuous parents will find the mean between
"acceptingness" and "committedness" – the mean between the
demands of PB and Sandel's stance on accepting the natural lottery
as a gift we should not intervene upon. The mean will look different
for parents conceiving naturally than for those conceiving through
IVF with the option of screening for disease and non-disease traits.

V Sex selection

Would a virtuous polity permit sex selection for non-medical purposes? This is very different than the question "Are all instances of
sex selection for non-medical purposes instances of parental vice?" I
won't attempt to answer this second question here, though I think it
is a fascinating one to consider, and my own view, for what it is worth,
is the answer is "no." But even if we thought that all instances of sex
selection (for non-medical reasons) were instances of vice, it doesn't
follow that a virtuous polity would necessarily seek to prohibit such
decisions. As Robert George notes in *Making Men Moral*, "Laws
cannot make men [*sic*] moral. Only men [*sic*] can do that; and they
can do it only by freely choosing to do the morally right thing to do
for the right reasons" (George 1993: 1). A virtuous polity will permit
its citizens to have the freedom to act wrongly; it does not prohibit all
moral and epistemic vices. Should a virtuous polity permit parents to
have the freedom to select the sex of embryos to implant, if parents
want such information?

Two general concerns are typically raised to justify limiting

restrictions on sex selection for non-medical purposes – concerns
of equality and concerns about the sex ratio. Concerns of equality
arise when the patriarchal practices in a society might put pressure
on couples to have a boy rather than a girl. In some Asian countries,
for example, having a girl can be considered a financial and cultural
liability because of the future dowry payments it will cost the parents.
The dilemma such countries face is that, if they permit abortion
(abortion became legal in India with the Medical Termination of
Pregnancy Act in 1971), and technologies such as amniocentesis and
ultrasound can determine the sex of the fetus, then parents might
intentionally abort female fetuses. And this has certainly been the
case in India, despite efforts to prohibit sex-selective abortion.

For virtue ethicists, *context* matters, and thus the issue of whether
to permit or prohibit PGD for sex selection purposes will be contingent upon the relevant background cultural factors, such as how
pervasive patriarchy is. The equality of women is clearly under threat
in a culture where sex selective abortion is commonly practiced. And
so there may be strong reasons, based on equality, for restricting the
use of PGD for sex selection purposes to avoid further exacerbating
inequality. However, combating such patriarchy by restricting access
to PGD may not be the most direct and effective way of addressing
these cultural inequalities.

The moral stakes involved in these issues will be different in a
society that has greater equality for women, such as Canada and the
United States. In these countries the societal pressures to have a male
child, or first male child, do not exist (at least not to the same extent).
Furthermore, not all instances of sex selection are sexist or threaten
equality. Some parents might want to pursue sex selection for family-
balancing purposes. Perhaps a couple already has two sons and would
really like to try for a third child and improve the odds of having a
daughter. Not every instance of sex selection will threaten equality. And
thus I believe a virtuous polity would consider intermediate regulatory
options between a complete ban and no restrictions at all. Permitting
sex selection for parents that already have children of one sex would
help limit insidious motives and permit aspirations of family balancing.
Or a policy that just prohibits sex selection for the first child would
have the same effect. Virtuous governance requires taking a purposeful
approach to reproductive freedom and finding a fair balance between
the demands to respect parental autonomy and equality.

Another concern that is often raised against sex selection is that it will distort the overall sex ratio between males and females. "The sex ratio at conception is unbiased, the proportion of males increases during the first trimester, and total female mortality during pregnancy exceeds total male mortality" (Orzack et al. 2015: E2102). This means that the natural sex ratio at birth is approximately 105 males born for 100 females. However, "females have greater resistance to disease throughout life and greater overall longevity" (Hesketh and Xing 2006: 13271) and males have higher mortality rates in accidents, war and violence. Thus, while there naturally tends to be a higher sex ratio at birth for males, differential sex mortality can offset this such that the population sex ratio is closer to 50 percent male and 50 percent female. But, once again, culture matters. For example, Hesketh and Xing studied the impact of sex-selective abortion on the sex ratio and concluded that:

> Largely as a result of this practice, there are now an estimated 80 million missing females in India and China alone. The large cohorts of "surplus" males now reaching adulthood are predominantly of low socioeconomic class, and concerns have been expressed that their lack of marriageability, and consequent marginalization in society, may lead to antisocial behavior and violence, threatening societal stability and security. (Ibid.)

Does the use of PGD for sex selection (rather than the case of sex-selective abortion) actually risk skewing the sex ratio at birth? Before answering that question, a virtuous polity would consider a number of factors. Firstly, it is important to remember that PGD can only be utilized when conceiving via IVF. Given the costs and burdens of IVF, this means only a small percentage of the population is likely to make use of it. And the pressures to select one sex over the other will be different in Canada than in India. The case for restricting sex selection is higher when there is a compelling empirical basis for thinking it would skew the sex ratio. Intermediate regulatory options, such as permitting it for family balancing, is something the virtuous polity will explore as the epistemic virtues prescribe paying attention to different salient facts and evidence. I think it is unlikely that a virtuous polity could necessarily rule out permitting any sex selection for the purposes of family balancing in countries that do not have oppressive patriarchal practices likely to skew the sex ratio.

The regulation of PGD, whether it be for the screening of disease traits or non-disease traits, raises a host of important moral concerns – concerns about the scope and limits of reproductive freedom, the interests of the child born from a process that involves their parents "selecting" their embryo, and broader societal concerns about equality and the sex ratio. These points effectively illustrate why it is important to have robust and informed societal discussion and debate on how to reasonably manage and balance the various interests at stake.

Discussion questions

1 The principle of "procreative beneficence" states: "Couples (or single reproducers) should select the child, of the possible children they could have, who is expected to have the best life, or at least as good a life as the others, based on the relevant, available information" (Savulescu 2001: 415). Discuss and debate the pros and cons of this principle.
2 Do you believe that parents should have the option to pursue sex selection via pre-implantation genetic diagnosis for family-balancing purposes? If you think such an option should be prohibited, what are your reasons?
3 Discuss and debate the four cases (A–D) of Tom and Mary with respect to the virtues of acceptingness and committedness. What conclusions do you arrive at concerning the virtues and vices of the different decisions made by the couple?

6
Aging Research and Longevity

I Introduction: global aging

People are, on average, living for much longer than in the past and are having fewer children. "Increasing longevity and declining fertility rates have been shifting the age distribution of populations in all industrialized countries toward older age groups" (Anderson and Hussey 2000: 191). Life expectancy at birth for the global population now surpasses seventy years (67.5 for males, 73.3 for females) (Wang et al. 2012) and is expected to rise to age eighty-one by the end of this century (United Nations 2011: xviii). "Globally, the number of persons aged 60 or over is expected to more than triple by 2100, increasing from 784 million in 2011 to 2 billion in 2050 and 2.8 billion in 2100" (ibid.: xvi). The year 2050 will mark a truly unique time in human history, for it will be the first time that the number of persons in the world age sixty or older will outnumber the number of children (up to age fourteen). The aging of the world's populations brings novel health challenges, as the chronic diseases of late life have now replaced infectious diseases as the leading causes of death. Age is a major risk factor for chronic pain, disability and disease.

The aging of human populations, or indeed of any species, is "unnatural." In nature life is typically, as Thomas Hobbes so eloquently phrased it in the seventeenth century, "nasty, brutish and short." Scientists have long been aware of the fact that the aging of any species is an *artifact* of human intervention. In the wild, it is rare for species to age. Mice, dogs and fish, for example, do not age because the extrinsic risks of the world are so high they typically die long before they reach the end of their natural lifespan. But as our pets they age and can live long enough to develop cancer because we feed and protect them, shielding them from threats such as starvation, infection and predation. The same is true for laboratory mice being studied for scientific research, as they live in a protected environment and thus they can survive long enough to experience such effects of senescence as cancer or cognitive decline – conditions that a mouse in the wild would not typically experience.

Peter Medawar, who won the Nobel Prize in Physiology or Medicine in 1960, described senescence as something "revealed and made manifest by the most unnatural experiment of prolonging human life by sheltering it from the hazards of its natural existence" (Medawar 1952: 13). "Humans, and the animals we choose to protect, are the only species in which large numbers experience ageing" (Hayflick 2000: 269). Without the benefits of the knowledge of hygiene and biomedicine, only a small percentage of people would live long enough to experience what occurs when our "biological warranty" (Carnes et al. 2003) period expires. That warranty, for humans, is estimated to be approximately seven decades.[1]

The aging of the world's populations is something that the field of bioethics has been largely ill-prepared to tackle, as the principles and theories typically championed did not take seriously the complexity of concerns raised by the prospect of millions of people living beyond the age of the human biological warranty period. Some (e.g. Callaghan 1987) have urged we should address the challenge of population aging by using age as a cut-off point for the most expensive medical technologies, while others have urged reform on issues pertaining to end of life decision-making, such as medically assisted death (Schüklenk et al. 2011).

A different strategy, urged by a vocal contingent of scientists – including Robert Butler (1927–2010) the first director of the National Institute of Aging in the United States – is to aspire to slow human aging and to consider this a major priority for public health (Butler et al. 2008; Kaeberlein et al. 2015). Aspiring to slow the aging process is distinct from the desire to treat or cure a specific disease of aging, such as cancer, heart disease or stroke. "The major focus of biomedical research has traditionally been the pathogenesis and treatment of individual diseases, particularly those with substantial effects on morbidity and mortality" (Kaeberlein et al. 2015: 1191). In chapter 3 I described how the focus on the causation of pathology could be construed as the defining intellectual commitment of what I called the paradigm of "negative biology." Prioritizing "negative biology" proved very effective in helping societies combat the plurality of threats posed by communicable diseases. But it has yielded a much smaller health dividend when applied to the chronic diseases that afflict the world's aging populations. The authors of a "Leading edge" commentary in *Cell* recently remarked:

> NIH-funded research is structured to address the major diseases driving morbidity and mortality. Interrogating and developing therapeutics for one disease at a time has often been productive. Will, however, the success of this approach be sustainable for the chronic aging diseases, such as neurodegenerative and metabolic syndromes, most cancers, and cardiovascular disease? (Kennedy 2014: 709)

A virtuous polity must exercise "adaptability of intellect," and the aging of the human species raises a complex array of opportunities and challenges for our societies, challenges we have never faced before. Unravelling the genetics of exceptional healthy aging is particularly important for the health prospects of today's aging populations.

II Positive biology and centenarians

In chapter 3, I introduced a distinction between the study of "negative" and "positive" biology. The former focuses primarily on the causation of pathology, whereas the latter seeks to understand and explain exceptional examples of desired phenotypes. Healthy aging is a great example of the importance of positive biology. But confusion can easily arise when discussing aging and a gerontological intervention, so a few preliminary definitions (e.g. What is aging?) and clarificatory points are necessary before turning to the ethical analysis of retarding aging.

Aging is a major risk factor for chronic disease. And chronic diseases such as cancer, heart disease and stroke are the leading causes of death in the world today. What role does genetics play in the way humans *biologically* age? When discussing longevity and aging it is important to bear in mind that aging, or *senescence*, refers to *biological* rather than chronological aging. Everything ages chronically at the same rate because chronological aging is a human construction. We keep track of the hours in a day and the days in a year, and thus the chronological age of my computer, my bike and "me" can all be classified into this schema. My laptop is now four years old, my bike seven, and I am forty-eight.

But the rate of *biological aging* (or senescence) is not something determined by the human artifice of time-keeping. Rather, it is determined by the evolutionary and life history of a species. Aging is "the progressive loss of function accompanied by decreasing fertility and increasing mortality with advancing age" (Kirkwood and Austad 2000: 233). As we become older, our bodies and minds become more vulnerable to disease and disability. In chapter 3 we discussed why, according to evolutionary biology, this is the case. The *disposable soma theory* (Kirkwood 1977; Kirkwood and Holliday 1979) maintains that biological aging occurs because natural selection favours a strategy in which reproduction is made a higher biological priority (in terms of the utilization of resources) than the somatic maintenance needed for indefinite survival. There is thus a real race between reproduction and death, and all the species alive today are (for now at least) winning this race. But for all the species that are now extinct, the race was lost.

Not all species age biologically at the same rate – indeed, even within a species there can be quite a wide variation in the rate (e.g. dogs). In *The Long Tomorrow* Michael Rose notes that many factors can influence the longevity of a species because they impact the force of evolution by natural selection. Size, for example, really does matter in nature.

> If a species lives longer in nature, the force of natural selection will be increased at later years. Larger organisms can reproduce at later ages because they are more likely to be alive then, so the force will remain high at later ages. This fosters selection of genes that will tend to keep the larger alive still longer. (Rose 2005: 64–5)

The bowhead whale, for example, which can grow to 20 meters in length and has a maximum lifespan exceeding 200 years, is an important species to study for longevity science, and its genome has been sequenced. These whales do not become sexually mature until after age twenty, and gestation takes over a year. This can be contrasted with the biology of the tiny field mouse. In the wild this mouse is vulnerable to many predators, and its winning strategy is to reproduce early in life, with a short gestation period and a large litter.

In addition to size, having wings also confers a longevity benefit. And the reason for this is that, with wings, a species is better able to escape predators. It can also fly vast distances to find food (which is typically scarce). Thus flying animals tend to live longer in the wild: "all else being equal, the force of natural selection favors the continued survival of flying animals more than those that can't fly. Thus evolution produces parrots that can easily live 60 years, if not longer, while rodents of the same size that don't fly die in six years" (Rose 2005: 65). And, finally, having a thick shell (like turtles do) helps a species live longer. "Having a thick shell is an evolutionary anti-aging device, because it reduces mortality and thereby increases the force of natural selection at later ages" (ibid.: 66).

While other species might grow exceptionally large, or have wings or a shell, what about us humans? How long can humans live? Jeanne Louise Calment, from France, died in 1997 at the age of 122, and she was the oldest person whose age has been verified by official documents. What was her secret? Did she eat a special diet or pursue an exceptionally rigorous exercise regime? No. Healthy aging is of course a complex phenotype, and as such both environment and genes are important. But in the case of the exceptionally long-lived (age 100 or more), genetics is extremely important. Having a centenarian sibling increases one's chances of survival to very old age (Perls et al. 1998). Furthermore, one recent study found that the offspring of long-lived parents had significantly lower prevalence of hypertension (by 23 percent), diabetes mellitus (by 50 percent), heart attacks (by 60 percent) and strokes (no events reported) than several age-matched control groups (Atzmon et al. 2004).

The maximum human lifespan is believed to be around 125 years (Weon and Je 2009). But, barring some major medical breakthrough in aging research, the odds of anyone breaking Jeanne Louise Calment's record are not very high, despite the large size of the global population and how safe our environments are today compared with those that existed when she was born in 1875. It is important to note that the prevalence of supercentenarians, and even centenarians, is very low. In the United States and other industrialized nations, centenarians occur at a prevalence rate of about one per 6,000. And supercentenarians occur at a rate of about one per 7 million.[2]

The longest-lived humans are an important biological puzzle to examine not simply because they live so long but because they typically experience a delay, and compression, of morbidity. For example, there are three different categories of centenarians – "delayers," "survivors" and "escapers" (Evert et al. 2007). The "delayers" are people who make it to 100 years with a delay of the onset of common age-associated illness. "Survivors" are people who were diagnosed with an illness before the age of eighty but survived for at least two more decades. And the third category of centenarians is "escapers" – people who escaped the most lethal diseases, such as heart disease, non-skin cancer and stroke.

To appreciate how unique these exceptionally long-lived persons are one must bear in mind the reality that most people will suffer multi-morbidity decades earlier. For example, multiple chronic conditions are evident in 62 percent of Americans over age sixty-five (Vogeli et al. 2007). Not only do the longest-lived humans experience more years of healthy life than the average person, but it has also been observed that supercentenarians (Andersen et al. 2012) and centenarians (Perls 1997) have a compression of morbidity at the end of life. Exceptionally long-lived individuals perhaps hold the key to the goal of delaying disease and frailty, as well as compressing the period of time persons must experience these maladies at the end of life. As such, the field of biogerontology (or geroscience), which studies the biology of aging, must be an important priority of a polity that faces the challenge of realizing the virtues of benevolence and justice for an aging population.

III Isn't exercise alone enough?

When we shift our focus from the exceptionally long-lived (age 100 or more), and the *maximum* lifespan, to the average lifespan of the majority of humans, we find that environment (rather than genes) becomes a much more significant factor in influencing a person's health prospects. Smoking, a sedentary lifestyle and a poor diet can all increase a person's risk of developing disease and dying prematurely (especially if an individual has not inherited the longevity genes that confer greater disease resistance in late life). However, environmental interventions alone will lead neither to adding decades of health nor to the kind of compression of morbidity found in those with exceptional longevity.

A helpful distinction to note is that between *primary* and *secondary* aging. The definition of biological aging noted earlier by Kirkwood and Austad (aging equals the progressive loss of function accompanied by decreasing fertility and increasing mortality with advancing age) refers to "primary aging." We all, regardless of lifestyle and environmental factors, experience the loss of function, decreased fertility and increased mortality in late life. Primary aging is inevitable. Eating vegetables and exercising, for example, does not alter primary aging. But these lifestyle modifications can of course improve one's health in late life, because they do have an effect on *secondary aging*. Secondary aging is the "deterioration in tissue structure and biological function that is secondary to disease processes and harmful environmental factors" (Holloszy and Fontana 2007: 709). Exercise can increase life expectancy by altering effects on secondary aging.

> Moderate and high physical activity levels have been demonstrated to lead to 1.3 and 3.7 years more in total life expectancy and 1.1 and 3.2 more years lived without cardiovascular disease, respectively, for men aged 50 years or older compared with those who maintained a low physical activity level. For women the differences were 1.5 and 3.5 years in total life expectancy and 1.3 and 3.3 more years lived free of cardiovascular disease, respectively. (Franco et al. 2005: 2355)

Biogerontology deals with the biology of primary aging, and an applied gerontological intervention would modify the rate of primary aging, which is very distinct from what happens with, for example, exercising and alterations with secondary aging.

What are the upper limits of an *average* lifespan? How high can we expect life expectancy at birth to rise? This is actually a very contested issue among demographers. There are two rival schools of thought on this issue (Couzin-Frankel 2011). One school of thought – let's call this "the past predicts the future" camp – bases estimates of increases in life expectancy at birth based on the increases in life expectancy of the past. The main proponent of this approach is James Vaupel, a demographer at the Max Planck Institute for Demographic Research. The second school of thought, represented by S. Jay Olshansky in the School of Public Health at the University of Illinois at Chicago, critiques the first for its neglect of the fact that "biology constrains math!." The gains in life expectancy made in the first half of the twentieth century were made primarily by saving the young. But, once you have saved the young, argues this second demographic perspective, you cannot save them again. Thus we can expect the gains to slow down as the age of life expectancy for a population gets higher. Increasing the life expectancy of a population from age forty-five to age fifty is very different than increasing its age from age eighty to eighty-five.

This second school of thought takes insights from comparative biology seriously when making predictions about the likely gains in life expectancy for humans. As such I believe it encompasses the "epistemic virtues" much more effectively than the first school of thought, as "the past predicts the future" approach bases its predictions on mathematic extrapolations without taking biology seriously. But biology constrains math.

Consider, for example, the average height for a population over time. If, a hundred years ago (in 1917), the average height of males was 5 foot 7, and today (in 2017) it is 5 foot 9 (an increase of 2 inches over a century), would it make sense to predict that, by the year 2617, the average male height would be 6 foot 9? No. Not unless there was some genetic intervention that altered the average height potential of humans. There is a biological limit to *average* human height, and, while environmental interventions such as improved childhood nutrition have increased the average height of people alive today compared with a hundred years ago, further improvements in nutrition will not see similar increases in our height. The past is not always a reliable predictor of the future, especially when those predictions ignore biology. One could argue the same is true for the average lifespan of humans.

The second school of thought, represented by Olshansky, which takes biology (especially evolutionary biology) seriously, estimates that, even if a population such as that of the United States complied with the perfect lifestyle in terms of exercise and diet, it would not likely achieve a life expectancy at birth beyond age eighty-five. Such an estimate is based on insights from demography (Olshansky et al. 1990) as well as comparative biology (Carnes et al. 2003), which compares interspecies mortality. Jeffrey Fries explains how estimates of the human lifespan are arrived at:

There are several methods of estimating the human life span. One may use the anthropological formulas, reconstruct an ideal survival curve from the tail of the present curve using the assumption that these individuals have been essentially free of disease, make extrapolations from the rectangularizing survival curve, or use estimates based on observed decline in organ reserve. All suggest an average life span of approximately 85 years, with a distribution which includes 99 percent of individuals between the ages of 70 and 100. (Fries 2005: 808)

If the average maximum lifespan is approximately eighty-five years of age, what should the aspirations of the medical sciences be, especially for polities with aging populations? One might respond that exercise and a healthy lifestyle ought to be the top and only priority. "We don't need to aspire to change people's genes!," the critic might contest. "We simply need to get people off the couch and eating better." Such lifestyle modifications can certainly yield health dividends for persons and populations. But the fact that the average maximum lifespan of humans is estimated to be around age eighty-five shows that such lifestyle alterations will not dramatically increase the number of *healthy* years people will live compared to the health benefits conferred by the longevity genes associated with exceptional healthy aging.

In "Confronting the boundaries of human longevity," Olshansky and his colleagues introduce the concept of "manufacturing survival time" to illustrate how the development of an aging intervention is very distinct from interventions that treat or prevent a specific disease (Olshansky et al. 1998). There are two general strategies for manufacturing the survival time of an individual and population: (1) reclaiming survival time by reducing *avoidable mortality* (e.g. death from starvation, malaria, appendicitis, infection, accident, violence, cancer, heart disease, stroke, and so forth) and (2) reclaiming survival time by *extending the biological warranty period*.

A virtuous polity would obviously pursue a robust strategy of achieving (1) – reducing mortality risks by preventing starvation, infectious disease, violence, etc. Benevolence and justice prescribe that the virtuous polity act in prudent ways to prevent avoidable mortality. However, (2) is extremely important to recognize once the extrinsic risks posed in (1) have been dramatically reduced. The risks in (1) will never be completely eliminated. There will always be some people who are killed in accidents, for example.

Intervening in the aging process itself promises to add a new strategy for manufacturing survival time by extending the biological warranty period of humans. An added benefit of extending the biological warranty period is that it does not simply aspire to add more years to life (as reclaiming survival time does); rather, it could add more life to years because it would extend the *healthspan* (rather than the *frailspan*).

A virtuous polity will consider the "salient facts" when contemplating how best to exercise the virtues of benevolence and justice. And one very relevant fact is the possibility of manufacturing survival time by extending the biological warranty period of humans. Many significant advances in biogerontology suggest that age retardation is now a *feasible* goal.

IV Intervening in the aging process

Insights into the biology of aging reveal that the rate of aging is malleable and influenced by a tradeoff between reproduction and longevity. The disposal soma theory predicts that a greater investment in longevity should come at a cost to reproductive fitness. And a series of studies support that conjecture. For example, one study compared the fertility rates of men and women with exceptional longevity (Tabatabaie et al. 2011). These individuals were young adults in the 1920s, before reliable methods of birth control were widely available. The study found that the exceptionally long lived (both males and females) had an average of 2.01 children versus 2.53 children for the control group. These differences in fertility were not related to gender or to education level. But there were developmental differences among the individuals with exceptional longevity. They tended to reach menarche a year later than average, have their first child three years later, and have their last child 2.5 years later than average.

Other studies have examined the impact of castration on the longevity of men, and they effectively illustrate the longevity/reproduction tradeoff. Castrated men residing in a mental hospital lived fourteen years longer than intact men in the same hospital (Hamilton and Mestler 1969). And, historically, Korean eunuchs had an incidence rate of centenarians at least 130 times higher than that in present-day developed countries (Kyung-Jin et al. 2012). Such findings support what the disposable theory predicts – that longevity comes at a cost to reproductive fitness.

Calorie restriction (CR) has been studied for decades in a variety of species (e.g. mice) and extends lifespan by altering the rate of biological aging. CR induces stress response pathways in organisms, which results in longer life by slowing the rate of molecular and cellular decline. What is truly exciting about CR is not that organisms simply live longer. Longer life is not necessarily desirable, especially if it is achieved by simply keeping an organism alive in a frail and incapacitated state (extending the *frailspan*). But CR does the opposite of this. It extends life by keeping an organism *healthy* for a longer period of time (extending the *healthspan*/biological warranty period). Since research in the 1930s, scientists have known "that rats and mice that are given about 40 percent less food than they would eat on their own live about 40 percent longer than do fully fed controls" (Miller 2002: 160). CR in rodents delays many of the chronic conditions associated with aging and thus can be considered an "anti-aging" intervention.

CR is too burdensome to be pursed as a gerontological intervention for human populations, but the prospect of developing a drug that mimics calorie restriction might be a viable way to safely and effectively retard aging. Two potential drug interventions that are now being extensively studied are those which activate the sirtuin genes (sirtuins are proteins that are activated by calorie restriction) and those which target a protein called TOR (*t*arget *o*f *r*apamycin). Rapamycin, for example, is a drug that was developed from soil on Easter Island. It is currently used to help prevent rejection in patients who have undergone organ transplant. But recent experiments have found that consuming rapamycin can extend lifespan, including in mammals. Perhaps the most significant study was published in *Nature* in 2009 (Harrison et al. 2009). In that study, mice that were already 600 days old (which is roughly equivalent to a sixty-year-old human) were fed rapamycin. This intervention increased the median and maximal lifespan of both males and females.

What if we could safely alter this longevity/reproduction tradeoff in humans, so that people could enjoy more years of healthy life? The process of evolution through natural selection is a blind, unintentional process. If humans possessed the knowledge to be able to purposefully alter the tradeoff, to promote more health in late life, would this be morally permissible? Indeed, should such an aspiration be considered morally *obligatory*?

V Benevolence and justice: the case in favour of age retardation

What would be the moral basis of the case for manufacturing the survival time of humans via age retardation? The basis, I would contend, is the same as the moral basis for manufacturing the survival time of humans by reducing preventable mortality risks. If it is virtuous to aspire to prevent specific causes of death – cancer, heart disease, stroke and Alzheimer's disease – then it also follows that it is virtuous to aspire to develop the knowledge and technology that permit us to do all those things *simultaneously*.

One way to illuminate how the moral interests behind age retardation can be equated with those behind the prevention of specific mortality risks is to undermine the distinction between "saving lives" and "extending lives."[3] Any intervention that can be described as "saving a person's life" can be redescribed as "extending their life." Saving a person from a fatal traffic accident, providing them with life-saving surgery or treating them for a disease is simply another way of saying that you extended their life. Benevolence requires a polity to aspire to prevent harm, including disease, disability and death. Preventing these harms also confers a benefit.

In his discussion of the duty to prevent bad occurrences, Peter Singer deliberately limits his principle to refer only to the "prevention of something bad" rather than suggesting the duty requires us to benefit others (1972: 238). But preventing death from starvation confers the *benefit of a longer life*. It would be inhumane to suggest that the duty applies only to the prevention of starvation up to the age of the "average lifespan," but that after that age humans should be permitted to starve because mitigating such a vulnerability would be "extending a life" rather than "saving a life." Older persons must eat too as they are also vulnerable to starvation. The same applies to the prevention of the chronic diseases and disabilities of late life. The virtues of benevolence and justice provide weighty reasons for aspiring to prevent the most common chronic diseases that afflict the world's aging populations.

One of the challenges faced by the scientific field of biogerontology is combating the prevalence of what aging researcher Richard Miller calls "gerontologiphobia" – "the irrational fear that aging research is a public menace bound to produce a world filled with nonproductive, chronically disabled, unhappy senior citizens consuming more resources than they produce" (Miller 2002: 170). Robert Butler, the first director of the NIH's National Institutes of Aging, coined the term "ageism" in 1969 (Butler 1969), which means a systematic stereotyping of, and discrimination against, people because they are old. Ageism is reflected in such colloquialisms for elders as "coot," "croon," "geezer," "hag," … "out to pasture," "over the hill," and "washed up" (Palmore 1999).

Studies[4] have demonstrated that multiple stereotypes of the older person exist, such as the "grandmotherly type." And such stereotypes can impede a polity's exercise of the moral virtues when it comes to cultivating support and understanding for the importance of a gerontological intervention. If the media portrays older persons as simply "grandparents," with no goals or aspirations beyond familial ones, then younger persons will form a constrained representation of what life is like as a senior, as well as the importance of a field of scientific research such as biogerontology, which could help promote better health in late life. Accurate media depictions of older persons – contributing to their communities in different ways, pursuing hobbies, maintaining a romantic and active sex life, contributing to their community, or pursuing their hobbies and spending time with friends – can help overcome the ageist attitudes that undermine support for biogerontology.

VI Three concerns, vs. objections, to life extension

A virtuous polity will always consider the pros and cons of pursuing different courses of action with respect to realizing the virtues of benevolence and justice. So far in this chapter I have detailed the case *in favour* of seeing age retardation as an integral element of the virtues of benevolence and justice. Benevolence requires that a polity aspire to prevent avoidable harms – harms to both individual persons and populations. Failing to pursue technological advancements that will help prevent and delay disease, disability and death in late life contravenes not only benevolence but also the virtue of justice. Neglecting or obstructing the development of a gerontological intervention fails to treat older persons fairly and impartially. And the just polity would eschew ageism. This point can be illustrated this way. Can we think of any age category in the human lifespan where one would oppose pursuing measures that prevented morbidity? Does anyone oppose reducing the morbidity and mortality risks of children? Or young adults? Or adults in mid-life? I assume the answer, for a virtuous polity, is clearly "no" in all these cases. Would that answer suddenly change once we consider humans in the *post-reproductive stage* of life? I think the answer is clearly "no!" In fact, the vast amount of dollars already invested in research into treatments for specific diseases shows us that we care about the health and survival of all people, young and old. To aspire to prevent disease and suffering when someone is a child or young adult, but not when they are a grandparent, would be unvirtuous.

However, there may indeed be legitimate reasons, grounded in the virtues of benevolence and justice as well as the epistemic virtues, for exercising caution and foresight when aspiring to extend the human lifespan via age retardation. These reasons do not, I shall argue, constitute *objections* to pursuing age retardation, but they ought to be taken seriously in terms of constituting *valid concerns* a virtuous polity ought to address when considering the aspiration of age retardation.

In *Beyond Humanity*, Allen Buchanan makes a useful distinction between a *concern* about biomedical enhancement and *an objection*. The former is merely a "con," a reason against it. But an *objection* to an enhancement is a much stronger claim. An objection is an "all-things-considered" judgement that an enhancement is undesirable because the cons outweigh any pros. As Buchanan (2011: 71) notes, "all objections are concerns, but not all concerns are objections."

This distinction can be helpfully illuminated by considering the case of exercise. Suppose Bill is overweight and has a sedentary lifestyle and a family history of high blood pressure. He knows he must start exercising regularly and lose weight. He decides to start riding his bike to work each day, a commute of approximately 20 km in total. However, Bill's wife has concerns about him commuting to work via bike. She worries that the distance might be too onerous and that he might be hit by a car in the busy urban district where he works.

The reasons Bill's wife brings up are legitimate concerns, but they do not constitute an objection. Why? To forfeit the benefits of engaging in more rigorous daily exercise would be folly for Bill because there are real health risks if he doesn't change his lifestyle. In other words, there are both reasons for and against his cycling to work each day. How ought Bill to navigate such a dilemma?

There are precautions Bill can take to address the concerns raised by his caring wife. Firstly, he should consult with his family doctor about the amount of strenuous exercise he should engage in given his current health. Perhaps the doctor will advise him to gradually build up his endurance before biking both ways to work each day. After two weeks of shorter rides (e.g. 5 km), Bill builds up to distances of 10 km, and eventually, 20 km.

Furthermore, Bill can reduce his risk of injury by wearing safety equipment (e.g. bike helmet), learning and following the rules of the road for riding a bike, and taking a route to work that has a bike lane. While he cannot eliminate *all* the potential risks of harm to himself, if he takes these reasonable precautions, his decision to bike to work can be the right decision. Even his protective wife, while still slightly worried about him being safe, can now agree that Bill's decision to bike each day is, "all-things-considered," the right course of action.

Pursuing the goal of age retardation presents us with a similar predicament. There are concerns that (1) tending to the health needs of a growing aging population might contravene the concern to help the young (what we will call "the priority of the young concern"), (2) that increasing the lifespan will exacerbate problems with a growing population (what we shall call "the concern about population size"), and (3) that an aging intervention might exacerbate existing inequalities if it is available only to the rich (what we call "the inaccessibility concern"). A virtuous polity will respond according to all three of these concerns. I detail these three concerns in the remaining three sections of this chapter.

VII The priority of the young concern

One common reaction people typically have to the prospect of developing an aging intervention is that it seems unfair, if not unjust, to suggest we should prioritize medical research that might benefit people in late life when there are so many pressing health problems in the world facing the young. Such a sentiment was expressed, for example, in a "Rapid response" to an article I published in the *British Medical Journal*. Commenting on my paper, which called for a greater investment in aging research, a general practitioner objected that it "seems faintly distasteful that we should be striving to extend life at the upper limit when so many millions have it snuffed out long before any 'natural' lifespan, and not because of the ageing process but because of infectious disease, malnutrition, inadequate hygiene, absence of basic midwifery care etc."[5]

This sentiment that the young have a moral priority is also expressed in what is called the "fair innings argument" (Harris 1985; Williams 1997). This argument maintains that the young should have some priority over the aged. Suppose, for example, a scenario where resources are so constrained that we have medical provisions to save only one life. However, two patients are in need of the limited life-saving medical treatment. The first patient is a young child, age five. The second patient is age seventy-five. If we had to choose which one of the patients to save, many would contend that the child should get the priority because they have not yet had the opportunity to live a full life. A virtuous agent would be moved, one might contend, by both benevolence and justice to aid the younger patient in this scenario.

It is one thing to contend that, *all-else-being-equal* (e.g. risk of death, likelihood of successfully intervening to prevent harm, cost of intervening, etc.), some priority should be accorded to aiding those who are "worse off." But this point must not be conflated with the view that no moral weight whatsoever should be accorded to the prevention of harms in late life. There are two important variables in the scenario described above that do not apply in the more general context of tackling the challenges of scarcity with respect to clinical medicine and public health in aging populations. Those variables are (1) in the two-patient scenario the situation described is a *zero-sum game*.[6] That is, there is only one medicine to offer, and it confers extra health to the younger patient or the older patient at the cost of the life of the other. And (2) in the two-patient scenario the numbers of young vs. older patients to be saved are *equal*. That is, either we save the one younger patient or the one older patient. The sentiment that the moral interests behind saving the young are significantly more weighty than saving the old is particularly compelling in this example because of these two variables.

Now let us consider more realistic scenarios where both of these variables no longer exist. In the first scenario a polity has varying numbers of persons at risk of morbidity. While there are children at risk of rare disorders and diseases, there are many individuals across the human lifespan at risk of different types of morbidity, *and* it has the resources to pursue preventative and therapeutic interventions that could help redress many of these risks and harms.

According to the Cancer Society of Canada (Canadian Cancer Society's Advisory Committee on Cancer Statistics 2017: 55), cancer deaths among children (aged up to fourteen years) accounted for less than 0.2 percent of all cancer deaths in Canada. "In 2017, almost 96% of cancer deaths in Canada will occur in people 50 years of age and older, with the median age of death estimated to be between 70 and 79 years for both sexes" (ibid.: 54). In the five years from 2008 to 2012, 595 children died of cancer out of a total of 219,540 deaths attributed to the disease (ibid.: 55). For every child that dies of cancer in Canada, there are hundreds of deaths among individuals over the age of seventy. Our individual thought experiment of saving one child versus one senior looks much more complex and challenging to resolve if the example were modified to saving one child vs. saving hundreds of people in late life. And in real life, unlike our zero-sum game example, we could potentially save the lives of both children *and* older adults at risk of cancer. To proclaim that we should be concerned with saving only the young would be to adopt a distorted moral lens that gives no moral weight at all to the interests of older persons. It is one thing to say that saving the young matters more, morally speaking; but it is quite another to say that saving the young is the only thing that matters morally.

The point of highlighting the disparity in cancer deaths between the young and the old is to suggest that the *numbers* of people suffering particular forms of morbidity ought to count in addition to the age of onset of the condition. Once one adds all the other morbidities that typically afflict mostly older persons, such as heart disease, stroke, etc., one can appreciate better the point that well-ordered science, for an aging polity and world, ought to aspire to delay and prevent the diseases of late life (in addition to aspiring to mitigate the vulnerabilities facing the young in the world). Furthermore, most young children alive today will live long enough to develop, and eventually die from, the chronic diseases associated with aging. So an aging intervention would benefit the vast majority of young people alive today by helping to protect them against chronic disease as they get older.

Critics of the argument I have deployed above might make an objection that is typically raised against a cost–benefit analysis (CBA) of priority-setting of scare resources, medical or otherwise. This objection takes issue with the moral weight CBA attributes to *aggregation*. In *Why Deliberative Democracy?*, Amy Gutmann and Dennis Thompson (2004: 17) provide an illustrative example of this worry. In the early 1990s the state of Oregon had to grapple with the problem of rationing medical resources for residents on Medicaid. The state's Health Commission created a lengthy list of conditions and treatments and ranked them according to a cost–benefit analysis. Treatments that conferred the greatest health benefit at the lowest cost ranked among the highest on the list, whereas those that were expensive but yielded a low return in terms of the health benefit were marked down. One counter-intuitive result of this CBA ranking, as Gutmann and Thompson highlight, was that capping a tooth ranked higher than an appendectomy. Capping a tooth was relatively inexpensive compared to an appendectomy, and thus you could benefit many more people by covering capped teeth over an appendectomy for far fewer people. And yet the latter was treatment for a much more serious medical condition. It seemed unfair to prioritize a less important medical procedure over a more important one simply because the numbers of people who would benefit from capped teeth was much larger than the number who would benefit from the more expensive appendectomy.

The example of prioritizing capping a tooth over an appendectomy is a compelling one because it demonstrates how the virtues of benevolence and justice require a more nuanced ethic than what is typically prescribed by CBA and its tendency to aggregate benefits in a way that leads to unfair outcomes. But, in the case of prioritizing the prevention, or even treatment, of late onset disease and disability, the stakes at risk are much more significant than capping a tooth.

In *What We Owe to Each Other* (1998), the contract theorist Thomas Scanlon explores how a contractualist could incorporate some room for aggregation without leading to the counter-intuitive consequences of utilitarianism or CBA.[7] And I think this same point applies to a virtue ethics endorsement of pursuing an aging intervention. Unlike utilitarianism, which aggregates, Scanlon's version of contractualism considers the individual standpoint of individual persons. And, from the individual standpoint of those who will suffer the diseases and afflictions of senescence (i.e. the vast majority of human beings), it would be irrational and unfair not to tackle aging itself in addition to tackling the diseases of aging. While Scanlon's contractualism rules out aggregation across lives, it does permit aggregation *within* each person's life. Scanlon permits numbers to count when the harm at issue is "morally relevant." And the harms of senescence are morally relevant (even though not equivalent) to early onset disease. Cancer cells, unlike normal cells, grow out of control and become invasive (spreading to other parts of the body). Whether it is early or late onset, cancer causes normal cells to grow out of control and impedes the development of normal cells. There are over two hundred different types of cancer, and, left untreated, cancer is fatal.

The Scanlonian insight that aggregation within each person's life is justified when the harm at issue is "morally relevant" lends more support to the perspective that the virtuous polity would treat age retardation as a pressing moral imperative (1) given where the scientific prospects of an aging intervention are, (2) given the demographics of the aging world, and (3) given the high prevalence of multi-morbidity in late life. Furthermore, it is worth emphasizing that, if an aging intervention simultaneously delayed most of the diseases associated with late life, this would be a much more cost-effective way of promoting health than trying to find treatments or cures for each specific disease of aging. So retarding aging could allow a polity to spend less on treating the diseases of aging and more on helping the young and/or on meeting other demands of benevolence and justice.

In *Just Health*, the bioethicist Normal Daniels dedicates a whole chapter to the issue of "Global ageing and intergenerational equity," where he defends what he calls the general Prudential Lifespan Account to health. This scheme involves treating people equally over their whole lives. Daniels argues that this account imposes two requirements that ensure our decisions concerning distributive allocations of healthcare resources and services are fair and impartial (Daniels 2008: 174).[8] Firstly, we must function behind a form of "veil of ignorance" and thus should pretend we do not know how old we actually are. And, secondly, we accept a distribution only if we are willing to live with the results at each stage of our lives. The Prudential Lifespan Account reveals why tackling aging itself should be more of a priority. The last years of most people's lives are the ones where they are most vulnerable to frailty, morbidity and mortality. Given the severity of the harms of senescence, it makes little sense not to invest aggressively in biogerontology.

The aspiration to retard aging can be defended on grounds of appeals to aggregation across lives, aggregation within lives, and the impartiality of Daniels's Prudential Lifespan Account. The "priority of the young" objection to intervening in aging is (arguably) more valid when it is raised against the project of "negative" (rather than "positive biology") – that is, when the objection is to the goal of trying to treat and cure each specific disease of aging. The fact of co-morbidity, coupled with the large numbers of older persons, means that more and more limited resources will need to be spent trying to reclaim potential survival time if a polity continues down the disease-specific approach to medical research. Unfortunately this leads to an extension of the human *frailspan*, as it leaves intact the "biological warranty period" humans have inherited from our evolutionary history but then invests enormous amounts of funding into research to prevent and treat all the specific diseases of aging. What advocates of an aging intervention are proposing instead is one single intervention that could simultaneously delay the onset of all age-related ailments. Richard Miller has estimated that the obstacles blocking the development of the hypothetical discipline of applied gerontology are 85 percent *political* and only 15 percent scientific (Miller 2002: 172). A virtuous polity would aim to overcome those political obstacles so that healthy aging could be a viable aspiration for the vast majority of humans, not simply those who have inherited "longevity genes."

VIII The concern about population size

A second common objection typically raised against the prospect of increasing the lifespan of humans via an aging intervention is the worry that this will cause an increase in human population, thus exacerbating pressing societal problems such as population density and environmental degradation. If people are living longer, they will consume more food and resources and contribute more carbon emissions.

Like the *priority of the young* concern, I believe the *population size objection* is actually a *concern* rather than an objection. As such, addressing the concern of population size can better prepare the virtuous polity to take the necessary steps now to try to minimize any potential adverse consequences likely to arise as a result of increasing the lifespan of humans.

Perhaps the first thing to note when assessing the issue of population growth is the fact that the world is a varied and diverse place. Some countries are plagued by famine, persistent conflict and high fertility rates, whereas other countries are economically prosperous, peaceful and have low fertility rates (below replacement levels). This means a virtue ethics analysis of the impact of slowing aging must be *context specific* – to consider the particular situation of any given polity.

Developed countries such as Canada, the UK and the United States have below replacement fertility levels and pursue immigration as an essential driver of economic growth, as having healthy, productive workers is vital for the labour market of a polity. A health intervention that extended the period of time citizens were healthy, and potentially working, is thus not necessarily a bad thing. In fact, healthy productive persons are arguably *the* most valuable resource for any polity. Unlike the aspiration to manufacture survival time beyond the biological warranty period by eliminating specific diseases of aging, the goal of age retardation has the potential to increase the *healthspan* rather than *frailspan*. As such, it could be a public health intervention that helps the world's aging populations remain economically viable by keeping people healthy longer and compressing (instead of extending) the period of morbidity at the end of life.

Furthermore, an aging intervention would be enormously beneficial in developing countries precisely because many vulnerable aging persons in those countries lack access to the benefits enjoyed by people living in richer countries (e.g. a pension and universal healthcare). Age retardation, like all other public health measures that help prevent the onset of disease and disability, would thus be beneficial not only to individuals and their families but also to the society at large.

However, as the critic might respond, more people means more food must be provided and more pollution will be created, thus a greater environmental toll on an already warming planet. Surely, our critic will contend, these "cons" of extending lifespan must be taken seriously? And I agree they should. But I do not believe a virtuous polity would eschew public health measures to address these problems. To see why, consider our attitudes about these other existing forms of "manufacturing survival time":

1 less war;
2 more exercise;
3 more smoking cessation;
4 less cancer;
5 better automobile design to reduce fatalities in accidents.

All of these measures, if realized, would increase the human population living on the planet. And yet I suspect no one raises the issue of overpopulation as an objection, or even concern, in any of these cases. No one declares "Let's have more car accidents, war or smoking to help keep the world's population size down!" And why not? Because having people die in car accidents, in war or from lung cancer are also very bad things. The appropriate response to concerns of population growth, in countries where it is a pressing problem, is to address high fertility rather than eschew health interventions and innovations. To propose solutions to climate change that involve undermining the goal of manufacturing survival time for humans is a vice rather than a virtue because it violates the demands of both benevolence and justice. Consider, for example, that the World Health Organization estimates that, under a base case socio-economic scenario, there will be approximately 250,000 additional deaths due to climate change per year between 2030 and 2050 (2014: 1). These are certainly worrying numbers, and action should be taken to try to prevent these additional deaths. However, the annual death toll of chronic disease *today* (rather than in thirty years), especially of those that occur in later life (over the age of sixty) are in the *millions* (rather than thousands), and this number will only increase this century. It is unreasonable and irrational to propose forfeiting health innovations that could help save the lives *of millions* of people *today* for the possible benefit of saving the lives of a few hundred thousand people thirty years from now.

The virtuous response to the problems of population density and climate change is not to criticize or object to longevity science (or any other public health measure) but, rather, to endorse the fair and *empirically demonstrated* measures which have been successful in curtailing population growth. The tried and tested solutions include education (especially of women), access to birth control, economic development, alterations in social attitudes towards the family and gender, etc. So a virtuous polity will be committed to the goals of both promoting healthy aging *and* addressing problems such as climate change. It is a mistake to think we must address only one and not the other. Pursuing both means encouraging innovation in the medical sciences and innovation in energy technologies. The former need not erode our commitment to the latter. However, serious attention and concern must be given to the potential impact extending the lifespan will have on already pressing societal issues such as the environment. And a virtuous polity would take the appropriate steps to minimize any adverse impacts that might arise as a result.

IX Equal access concern

Many critics of genetic enhancements raise the objection that they will exacerbate the inequalities that already exist, and persist, in today's societies. Walter Glannon, for example, in *Genes and Future People*, goes so far as to say that genetic enhancements should be impermissible because unequal access to such interventions could undermine our belief in the importance of the fundamental equality of all people. He argues, "Allowing inequalities in access to and possession of competitive goods at any level of functioning or welfare might erode this basis and the ideas of harmony and stability that rest on it" (Glannon 2001: 100).

The issue of who would have access to an aging intervention, both domestically and globally, certainly does raise valid concerns that a virtuous polity ought to anticipate and address. But this does not mean this worry constitutes an *objection* to pursuing, indeed prioritizing, such research. Consider, for example, the fact that most interventions that have manufactured human survival – ranging from sanitation and immunizations to advances in medical knowledge – are not equally accessible to all humans living in the world today.

This reality does not lead us to conclude that the virtuous thing to do is forfeit such technologies for everyone. Instead, one can make a compelling case that justice prescribes that measures should be taken to ensure that the poor also have access to things such as clear drinking water, immunizations, etc. The virtuous response to the inequality concern is to support policy initiatives that help reduce, if not eliminate, such inequalities. The goal is not to achieve greater equality by levelling down (so that no one has access to health innovations) but, rather, to achieve greater equality by leveling up.

If a drug that mimics the effects of calorie restriction is developed, and it proves to be a safe and effective way of increasing the healthspan of humans, then it ought to be considered a "medical necessity" and made accessible to all by a benevolent and just polity. Rather than eschewing an aging intervention, those who raise the "equal access" concern should support the creation and fair diffusion of such an innovation. Doing so might even prove to be a much more feasible and effective way of redressing inequality than trying to achieve equality of access to each healthcare provision being pursued or offered by the paradigm of "negative biology." By making an aging intervention an integral part of its approach to public health in the twenty-first century, the virtuous polity is less likely to exacerbate the injustices that might arise if aging research is obstructed because concerns about intervening in aging were conflated with objections, or if such medical research is simply left to a market regulation and thus only available to those who can afford to pay for it.

Discussion questions

1 Should we aspire to intervene in the aging process of humans? Why/ why not? Consider the concerns of the priority of the young, population growth, and access to any aging intervention. Are any of those concerns pressing enough to warrant an *objection* to pursuing an aging intervention?

2 Is there a meaningful distinction between "saving" and "extending" a person's life? Do the virtues of benevolence and justice apply only to the former and not to the latter? Why/why not?

3 "What really matters is increasing the human *healthspan*, not the number of years a person can remain alive in a frail condition manag-

ing multi-morbidity. We should aspire to add more life to our years, not more years to our life." Do you agree or disagree with this statement? And what significance does this have for the way we undertake medical research?

7

Happiness, Memory and Behaviour

I The greatest happiness of the greatest number?

Throughout this book I have utilized a general "virtue ethics" normative framework for assessing some of the ethical and societal-level
implications of advances in our understanding of human genetics – a
framework which emphasizes the importance of our exercising both
moral and intellectual virtue to capitalize on the potential benefits
(while avoiding the potential risks of harm) of genetic knowledge
and novel biomedical interventions. One normative theory which
equates moral decision-making with actions that bring about the
best expected consequences is *consequentialism*. Virtue ethics and
consequentialism are typically presented as rival normative theories.
Not only do these two moral traditions disagree about what makes an
action morally right (e.g. it brings about the best consequences vs. it is
the action a virtuous agent would choose), they often have competing
understandings of what constitutes human wellbeing or happiness. In
this chapter we will explore some of the novel challenges presented
by advances in our understanding of the role genes play in happiness,
memory and human nature for both consequentialists and virtue
ethicists.

Consequentialism is the moral theory which maintains that the
morally right action or policy is that which creates the best consequences. The term "best consequences" has been interpreted
differently by different consequentialists. The late eighteenth-/early

nineteenth-century philosopher Jeremy Bentham (1748–1832), the founder of the consequentialist position known as "utilitarianism," argued that "the greatest happiness of the greatest number" constituted the best consequences. Bentham defended a *hedonistic* account of wellbeing, one where the best consequences could be equated with maximizing the sensations of pleasure and minimizing pain.

Bentham is perhaps most famous for his "calculus of happiness," a hedonistic formula for determining which actions to pursue in order to maximize pleasure and minimize pain. The calculus consists of the following seven elements:

1 the intensity of the pleasure;
2 the duration of the pleasure;
3 the certainty of the pleasure;
4 remoteness (nearness in time);
5 fecundity (likelihood that it will be followed by sensations of the same kind);
6 purity (likelihood that it will be followed by opposite sensations);
7 extent (the number of persons to whom it extends).

For example, if I am making a decision that will largely impact just the pleasure of myself (so the "extent" of persons is just one – namely, me) – such as spending my Friday evening reading more philosophy or watching my favorite TV show – I would have to consider, for instance, not only the difference in intensity (e.g. perhaps watching the show is a more intense pleasure than reading) but also the duration of the pleasures (one might last for years whereas the other just a few hours), the probability of one activity leading to other pleasures or pain further down the road, etc. And when applied to societal-level decisions, involving the economy or climate change, we see how the issues get extremely complex and contentious.

While there are few defenders today of Bentham's original calculus of happiness, his theoretical framework did inspire cost–benefit analysis (CBA), a very influential tool in public policy. CBA requires regulators to consider the costs and benefits of different options and determine which one creates the best overall benefits at the lowest cost.

Bentham's hedonistic moral theory is predicated upon a simplified account of human psychology – an account that reduces our mental life to either pleasurable or painful states. In *An Introduction to the Principles of Morals and Legislation*, Bentham famously remarked: "Nature has placed mankind under the governance of two sovereign masters, pain and pleasure. It is for them alone to point out what we ought to do, as well as to determine what we shall do" (cited in Wooton 2008: 585). But the process of evolution by natural selection yielded mental faculties that are much more complex than the picture presumed by Bentham, giving us reason to be skeptical that sage normative prescriptions can be gleaned from the conjecture that it is our nature to be "hedonic maximizers." Evolutionary biologists provide a much more expansive and complex account of our mental life:

> Our evolved natures should be treated with respect, but not with deference. We did not evolve to be happy: rather we evolved to be happy, sad, miserable, angry, anxious, and depressed, as the mood takes us. We evolved to love and to hate, and to care and be callous. Our emotions are the carrots and sticks that our genes use to persuade us to achieve their ends. But their ends need not be our ends. Goodness and happiness may be goals attainable only by hoodwinking our genes. (Stearns et al. 2008: 13)

Despite the flawed understanding of human nature upon which hedonism is predicated, over the past two decades or so the "science of happiness," known as "positive psychology," has matured and yielded a number of significant practical prescriptions concerning how individuals, and societies, could increase our levels of wellbeing. And thus a serious examination of the role genes play in our happiness is certainly warranted.

Martin Seligman (2002), a pioneer in the field of positive psychology, distinguishes different kinds and levels of happiness. Hedonists who pursue the immediate positive feelings – such as the pleasure of a food they enjoy or a compliment – seek the *momentary* happiness of what Seligman calls "the pleasant life." But these pleasures fade quickly and thus do not have a lasting impact on the subjective wellbeing of people. Enduring happiness, the kind we enjoy when we live the truly "excellent life," is realized when we lead a *meaningful life*. After spending years of studying what makes people happy, Seligman remarks:

> What does Positive Psychology tell us about finding purpose in life, about leading a meaningful life beyond the good life? I am not sopho-

moric enough to put forward a complete theory of meaning, but I do know that it consists in attachment to something larger, and the larger the entity to which you attach yourself, the more meaning in your life. (2002: 14)

A concern with the "excellent life" is, at least historically, the primary focus of the virtue ethics tradition, especially for Aristotle. And this can be contrasted with Bentham's concern for maximizing the presence of positive affect (e.g. the subjective experience of positive emotions such as joy and interest) and the absence of negative affect.

The psychology of happiness is a fascinating and complex field of study that we cannot delve into in the amount of detail necessary to do it justice. My goal in this book is not to champion a particular account of happiness as the ultimate normative guide for our deliberations about all matters pertaining to advances in human genetics. Instead, my primary focus, at least till this stage, has been on the practical guidance that invoking moral and intellectual virtue and vice can offer us. But in this chapter I wish to make a basic, rudimentary distinction between the hedonic account of happiness championed by Bentham and an Aristotelian-inspired account of happiness understood as a "way of life." Deci and Ryan argue that "the two approaches to well-being – namely, hedonism and eudaimonism – are founded on different views of human nature" (2008: 3). For Aristotle, *eudaimonia* (happiness) is the highest cultivation of our character. Our highest goal is "to engage in an activity of the soul – or rather, the rational parts of the soul – in accordance with areté ('virtue', 'excellence')" (Kraut 2002: 70). Today many psychologists invoke an Aristotelian-inspired account of wellbeing, one that stands in contrast to the limited account of hedonism. "Eudaimonic well-being also involves pleasure but emphasizes meaningfulness and growth – a more enduring sort of happiness" (Bauer et al. 2008: 83). The emphasis on meaningfulness and growth, rather than just on positive and negative affect, assumes a conception of humans where *narrative identity* – that is, the story we tell ourselves about our lives – is an important determinant of our happiness or *eudaimonia*. Just to be clear, like Seligman, I am not sophomoric enough to put forth a complete characterization of what constitutes a meaningful life. I do not think trying to do so would serve this introductory genetic ethics book well. Instead, I want to suggest that research on the genetics of happiness, of both the hedonic and eudaimonic varieties, raises novel and intriguing ethical and societal issues worth addressing seriously. And my comments and arguments in this chapter aim to help spark such deliberations and debates rather than offer conclusive insights on the issues.

Daniel Kahneman and Jason Riis (2005) provide a helpful way of understanding the difference between the hedonic and eudaimonic accounts of happiness. They do not see them as stemming from two accounts of human nature; rather, they focus on two different perspectives on life – the "experiencing self" and the "remembering self." My "experiencing self" is the self that can report on how I feel at this particular moment in time, having just finished exercising, or during eating my favorite meal, or watching my favorite TV show, etc. My "remembering" or "evaluating self" self is the self that reports on my life events – such as the holiday I took four years ago, or my experience of being a father when my children were infants, or my time as a PhD student at university over twenty years ago. Both of these elements of our happiness and wellbeing are important. And we are only beginning to understand the role genes and environment play for both selves.

Perhaps the clearest illustration of the limitations of the hedonic account of wellbeing is the interesting role of wealth in human happiness. People often assume that becoming much richer would make them much happier. But surveys in many countries conducted over decades indicate that, on average, reported global judgements of life satisfaction or happiness have not changed much over the last four decades, in spite of large increases in real income per capita (Kahneman et al. 2006: 1908). Wealth is much more important to happiness when we are talking about assets below a certain minimum threshold (e.g. poverty). Not having money for housing or food can certainly impact a person's level of happiness. But a study in the United States concluded that emotional wellbeing – the frequency and intensity of experiences of joy, stress, sadness, anger, and affection that make one's life pleasant or unpleasant – did not progress beyond an annual income of about $75,000 (Kahneman and Deaton 2010). And yet many people making more than this baseline of "happiness income" still pursue the imperative to get more resources and "stuff." Why?

Culture no doubt explains part of the reason. The capitalist culture and media, etc., portray affluence and the possession of material resources as ideals to strive for. But part of the reason these cultural pressures are so effective is because they tap some primal biological imperatives. The desire for individuals and collectives to maximize material resources no doubt served an important evolutionary function when scarcity of resources was a fact of life and threatened our survival. This was our reality for over 99 percent of human history and still is a reality for the poorest regions of the world. But these basic (adaptive) instincts can jeopardize our ability to realize flourishing lives when the environmental circumstances we face are very different than those of our evolutionary past. The problem of obesity is a great illustration of this. Striving to fulfill the imperative to maximize calorie intake is a recipe for a premature death in the modern world of developed countries that have an abundance of cheap, high calorie foods.

Money can improve a person's happiness if it is spent in the right way – for example, if it is spent on experiences rather than material things (Pchelin and Howell 2014) or on benefiting other people. Elizabeth Dunn et al. (2008) found that, when individuals spend more money on *prosocial* goals such as charity, they actually experience greater happiness than when they spend money on themselves. Such findings illustrate the diverse natures humans have inherited from our Darwinian past. On the one hand, we may be inclined to seek to maximize our material prosperity, which was prudent when we faced, as humans did for most of our species' history (and millions of people still do today), difficult external conditions such as scarcity of resources. And yet, on the other hand, we are also inclined to cooperate and aid others, even at a great cost to ourselves. The latter is what we would expect if humans really are, at least at some level, the social and political animals that Aristotle claimed we are rather than selfish agents seeking only to maximize our hedonic pleasures.

The influence of heredity on happiness is the subject of much debate, as is the prospect that we could, one day, develop interventions to consciously engineer humans to be "happier." In "Happiness is a stochastic phenomenon," Lykken and Tellegen (1996) measured the subjective wellbeing of several thousand middle-aged twins. This study examined a person's "experiencing self" and asked participants questions such as: "Taking the good with the bad, how happy and contented are you on the average now, compared with other people?" One might expect obvious things, such as a person's wealth and income, educational attainment or marital status, would explain most of the variations in reported levels of happiness. But Lykken and Tellege's findings were very surprising.

Educational attainment accounted for less than 2 percent of the variance in WB [wellbeing] for women and less than 1 percent of the variance for men (Lykken and Tellegen 1996: 187). And "positive mood states are not much more frequent or intense for people with high social status or wealth and that people at the lower end of the social ladder are only slightly more vulnerable to negative mood states" (ibid.: 188). The standard factors that we expect to account for differences in wellbeing – education, money, marital status and religious commitment – accounted for a variance of only 3 percent. Following up with a smaller set of twins at intervals of 4.5 years and ten years, the study concluded that the heritability of the stable component of subjective wellbeing approaches 80 percent. The authors even remarked that "it may be that trying to be happier is as futile as trying to be taller and therefore is counterproductive" (ibid.: 189). This estimate that genes account for 80 percent of a person's happiness is contentious and debated. In *The Politics of Happiness* Derek Bok surveys the findings on the genetics of happiness and concludes that "the consensus of researchers today is that heredity probably accounts for at most 50 percent of one's happiness level but that the other 50 percent is determined by events and circumstances and deliberate choices" (2010: 189).

The strong influence genes have on our happiness is also illustrated by recent genome-wide association studies of three phenotypes: subjective wellbeing, depressive symptoms and neuroticism (Okbay et al. 2016). If genetics accounts for so much of the variation in levels of happiness in our society, then the prospect of developing interventions that could intentionally alter our genes or gene expression are indeed profound! If the just society is one committed to "Life, liberty and the pursuit of happiness," and this last value is so significantly influenced by our genes, it would seem that a robust societal-level aspiration to modulate the "genetic inequality" for happiness should be undertaken. Why should some people enjoy fewer positive emotions (e.g. gratitude, elation, amusement, etc.) and more negative emotions (e.g. anxiety, fear, anger, etc.) than others simply because of the genes with which they were born? Why do our societies continue to focus so much on socio-economic differences (such as wealth and educational attainment) when the genetic lottery of life also has a profound influence on our levels of happiness? These are valid questions to raise and debate today.

Advances in our understanding of how our genes influence our wellbeing raise significant ethical and social questions. Would increasing our level of happiness (both positive emotions and *eudaimonia*) via genetic engineering be morally desirable? Or would it compromise our ability to lead "authentic" human lives rather than just hedonically pleasant lives? I believe a genetic intervention that could alter our potential for happiness would indeed constitute one of the most significant potential interventions we could develop, even if we are not committed to utilitarianism or virtue ethics. However, much would depend on the details of how it could actually do this (e.g. permitting us to enjoy more positive emotions, less unhealthy anxiety and stress, more humility, less irrationality and bias, etc.). I remain skeptical that appeals to the general principle "the greatest happiness for the greatest number" will be sufficient to ethically regulate any such intervention. To appreciate the myriad of considerations that are likely to arise in this context, we will now examine, in greater detail, the prospect of developing new ways of modulating human memory to help enhance the emotional resilience of soldiers to prevent post-traumatic stress disorder.

II Insulating soldiers from the emotional costs of war – virtue or vice?

Rapid advances in the biomedical sciences raise the prospect that it may be possible to go beyond the traditional therapeutic aims of medicine (e.g. treatment of disease) to actually "enhance" the biology of humans. In *Beyond Humanity*, the bioethicist Allen Buchanan defines biological enhancement as follows: "a deliberate intervention, applying biomedical science, which aims to improve an existing capacity that most or all normal human beings typically have, or to create a new capacity, by acting directly on the body or brain" (2011: 23). From increasing the human lifespan to improving strength, intelligence and memory, the topic of *human enhancement* raises a host of ethical and societal concerns.

The prospect of enhancing human performance is particularly important, and controversial, in two domains where maintaining, indeed pushing, the boundaries of peak human performance is crucial to the activity – sports and the military. Athletes need to remain in the best physical condition possible if they hope to be competitive on the world stage. And this pressure to be the fastest and/or the strongest has led to performance-enhancing substance scandals in sports, from cycling and baseball to weightlifting and sprinting. The names Ben Johnson, Lance Armstrong and Barry Bonds will be remembered more for their use of performance-enhancing substances than for their athletic prowess.

Enhancement within the military also raises a plethora of complex ethical issues. Countries want their military to be effective so that, should their service men and women be called upon to engage in risky military operations, they can successfully achieve their military objectives with minimal risk of suffering injury and loss of life.

Critics of human enhancements often object that they are "unnatural" (Kourany 2014: 985). Applied to the prospect of enhancing the emotional resilience of soldiers through new memory-altering drugs (hereafter referred to as MADs), this "unnatural" objection often invokes an overly idealized account of the role memory plays, and should play, in human identity and wellbeing. For example, the President's Council on Bioethics report *Beyond Therapy* claims that memory-blunting interventions risk "falsifying our perception and understanding of the world" (2003: 228). As such, memory-blunting technologies could threaten the ability of the modified person living an *authentic* human life.

In the next few sections of this chapter I will argue that this line of objection to MADs is predicated upon a misunderstanding of how humans already modify, edit and suppress memories in order to reduce emotional discomfort and pain. Memory modification, whether unconscious or conscious, is an integral part of the "psychological immune system." Gilbert et al. (1998) describe this immune system as the brain's ability to protect us from gloom. To help us cope with the adversity we inevitably face in a hostile world, our mind can be very selective about the information it retains, distorts and omits. As such, memory modifications can be both *adaptive* (that is, positive to the individual) and *maladaptive* (negative to the individual). Whether or not any potential MAD is, *all-things-considered*, beneficial to a soldier (and thus a true "enhancement") will depend on the specifics of how it impacts their emotional resilience. A drug that helps insulate against post-traumatic stress disorder (PTSD) by altering the neuroplasticity of the brain so that remote traumas can be more effectively treated with cognitive behavioural therapy is a good example of such a potential intervention.

In the following section I begin to unravel the "enhancements are unnatural" objection by pointing out that the stressors of contemporary warfare are also "unnatural." I then consider the feasibility of developing a MAD to help reduce the emotional toll of warfare on soldiers by detailing a recent study on dulling fearful memories in mice by extending the period of time "recent" memories remain recent before becoming "remote" (and thus less receptive to behavioural therapy). The objection that memory modifications make us "less than human" or "inauthentic" ignores the reality that the "psychological immune system" already involves memory modifications, many of which are unconscious. This means MADs should not necessarily be ruled out of hand as inherently problematic simply because they alter our memories. If they help enhance the emotional resilience of soldiers in a way that is conducive, rather than corrosive, to their wellbeing, then they could be considered as *morally obligatory* interventions to provide to those at high risk of witnessing traumatic events.

However, as I detail in section VI, there is another moral duty that should not be circumvented by a duty to enhance the emotional resilience of soldiers. And that duty (which derives from the duty of benevolence) is to seek, as far as is possible and reasonable, to avoid placing soldiers (or civilians) in situations that will exacerbate the occurrence of traumatic events in the first place. In other words, the utilization of MADs should not lower the high moral threshold for justified military intervention. Peace is a cheap and effective way of preventing the traumas of conflict! Peace should be the ultimate, long-term solution to the problems of reducing the emotional trauma of conflict. But, in the non-ideal world, sometimes military conflict is necessary to avoid human traumas on a larger scale. And, in that non-ideal context of the contemporary world, I do not believe the "MADs are unnatural" objection is a persuasive argument against the development of safe and effective MADs.

III Modern warfare creates "unnatural" stressors

In 2011 Canada ended its 9.5 year military operation in Afghanistan. It is estimated that, of the 25,000 to 35,000 military members released from the Canadian forces between 2011 and 2016, at least 2,750 individuals can be expected to suffer from a severe form of post-traumatic stress disorder, and at least 5,900 will suffer from a mental health problem diagnosed by a health professional.[1]

The emotional toll of participating in military operations can be significant. And while the armour soldiers wear when engaging in conflict has dramatically improved in the past century, thus helping to protect their bodies from the external threats of bullets and explosives, the human brain which helps protect us from some of the emotional costs of trauma has not advanced. Hence the large number of cases of PTSD suffered by soldiers. But the prospect of developing MADs that enhance the psychological immune system could help make soldiers more emotionally resilient to the emotional toll of war and conflict.

Like the case of MADs, critics of an enhancement that alters the aging process of humans (which we examined in greater detail in the last chapter) often claim such an intervention is "unnatural."[2] One way to illustrate the point that the "enhancements are unnatural" worry is not compelling with respect to an aging intervention is to emphasize the point that the aging of human populations itself is "unnatural." If the aging of our populations is "unnatural," then an "unnatural" solution might be an appropriate response!

Noting how the reduction in the extrinsic risks of morbidity and mortality has created the conditions necessary for the chronic diseases of late life to become prevalent helps undermine the "an aging intervention is unnatural" objection to developing a gerontological intervention. We humans created the conditions for population aging. And if we hope to reduce the suffering that will arise from humans surviving past the seventh, and perhaps even eighth decade of life, we should support an aging intervention that delays the onset of chronic disease and disability.

This very same point can be emphasized to support the case of MADs. While warfare has always been a part of human life, there are many unique stressors of modern warfare that challenge the psychological immune system in ways that it did not evolve to deal with. The psychological immune system humans have today was shaped over the 200,000 years or so[3] of our species' evolutionary history. While violence and conflict was a reality of life for humans, the emotional trauma caused by modern warfare is very new and distinct. "Most hostile intergroup contact among huntergatherers was probably ongoing or intermittent, with occasional casualties, more akin to boundary conflicts among chimpanzees than to the pitched battles of modern warfare" (Bowles 2009: 1294). The emotional trauma a tribal warrior might face when killing the enemy of a neighbouring tribe in hand-to-hand combat, or with a bow or spear, is very different from the emotional trauma of witnessing the carnage caused by an m2 machine gun, an apache helicopter or a cruise missile.

The sheer scale of death and injury that the weaponry of modern warfare can inflict exceeds anything that the human psychological immune system evolved to deal with. Furthermore, there are many unique aspects of the nature of modern warfare that can add novel stress on soldiers. Intergroup conflict, in the distant past, would have had a very clear rationale – *group survival*. LeBlanc (2014: 29), for example, argues that resource stress and carrying capacity limitations are the primary cause of forager warfare. Combatants that killed invaders attacking their tribe would have a clear rational for why they had to take the life of others – *to protect their kin*. This is a very compelling justification for committing violence against others. Even offensive campaigns to take scarce resources from neighbouring tribes would have been rationalized as necessary for the survival of the group.

Contrast this with the strained rationalization a soldier who experiences trauma in a campaign such as Vietnam or Afghanistan might face. They might return home to mass public protests opposing the military intervention to which they contributed. Their compatriots might characterize their actions as evil, barbaric, etc. These societal pressures will add yet more strain on the psychological immune system of soldiers struggling to make sense of what happened.

The conditions of modern warfare contain many unique stressors that were absent from the evolutionary past. Thus one might contend that the psychological immune system should be updated, improved to deal with the stressors of modern warfare. If MADs are "unnatural," one might retort that the stressors of modern warfare are also "unnatural." And this deflates the strength of the "enhancements are unnatural" objection.

While I think this type of rebuttal to the "MADs are unnatural"-type objection is effective, it does raise a potential problem concerning the duty to minimize the risk of soldiers being exposed to traumatic events in the first place. And this is especially relevant when considering cases where there is widespread opposition to the use of military force. Such cases are typically ones where the moral justification for using military force is fiercely contested. Utilizing MADs should not lower the moral threshold for justified military intervention. This point is addressed in section VI of this chapter.

IV MADs: science fiction or the science of tomorrow?

"Memory in the biological sense is best understood as the systems underlying our capacity for retaining, storing and recalling experiences" (Liao and Sandberg 2008: 86). Preventing or treating PTSD by subduing fearful or painful memories via novel pharmaceuticals raises a host of interesting questions. For example, what constitutes "excellent" memory (e.g. perfect recall, remembering only pleasant experiences, etc.)? Would an intervention that subdued the traumatic memories of combatants compromise their ability to live "authentic" lives? Would it make soldiers "less human"? Or perhaps such interventions ought to be conceived of simply as an extension of more conventional therapeutic techniques, such as the narrative changes elucidated by psychotherapy or cognitive behavioural therapy?

Ethical debates surrounding any biomedical enhancement tend to get highly emotive, especially when the potential technology under question is decidedly speculative. For example, consider an intervention that increased the human lifespan. If such an intervention is stated very simply as "adding twenty years to life expectancy," it is not very clear what, exactly, this means in terms of the quality of life it might confer. If the twenty-year increase is achieved simply by expanding the *frailspan* – that is, the period of time people live with multi-morbidity and disability – that would not necessarily be a desirable outcome. However, if the extra twenty years could be added to the human *healthspan*, so the intervention in question delays the chronic conditions of aging and perhaps even compresses the period of morbidity at the end of life, that is very different. Assessing the potential pros and cons of a potential "anti-aging" enhancement comes down to the specifics of what is gained, both for individuals and collectively as a society. Qualitative considerations matter, not just quantitative ones (i.e. the number of years alive).

I believe the same points apply to any potential MAD that improves the emotional resilience of humans. Would such an intervention improve the overall quality of life of people, so they can enjoy more positive emotions, cope better with the adversity life inevitably throws their way, etc.? To assess what is required by moral and epistemic virtue with respect to enhancements, it can be helpful to focus on technological interventions that have some close analogy to existing enhancing technologies. So, for an aging intervention, it is helpful to focus the ethical debate on a drug that mimics the effects calorie restriction has on slowing biological aging or a drug that activates the "longevity genes" which the longest-lived humans already have activated. We have already explored these points in chapter 6.

Like an ethical analysis of an aging intervention, any provisional moral conclusions that arise from an ethical analysis of MADs will largely depend upon many relevant facts. One could invoke highly imaginative interventions that allow humans simply to pop a pill and erase any selected memory – from the memory of a bad hair day to an embarrassing faux pas and toxic friendships or expired romantic relationships. These are the kind of scenarios that lead the President's Council on Bioethics to conclude that MADs could pose a significant threat to living an authentic human life. If used to blunt all the emotional unease we face, it will not actually *enhance* us because it will deny us the ability to learn how to cope and grow to deal better with different types of adversity. So if MADs were pursued simply to minimize or eliminate negative affect, at a cost to constitutive elements of our *eudaimonia*, then they could be detrimental to the realization of a meaningful life.

When it comes to a potential memory-altering drug, just like any potential aging intervention, the devil is in the details concerning whether or not, *all-things-considered*, the so-called enhancing intervention in question actually leaves a person better off. The case for seeing some potential MADs as an actual "enhancement" is more compelling, I shall now argue, when it is understood as enhancing our psychological immune system in ways that facilitate eudaimonic wellbeing. As such, MADs could be prescribed to soldiers to reduce the emotional toll of war by helping them to improve the efficacy of such traditional therapies as cognitive behavioural therapy. A recent study on fear in mice published in *Cell* is a great example of this and will be the focus of my ethical analysis of improving the emotional resilience of soldiers.

In "Epigenetic priming of memory updating during reconsolidation to attenuate remote fear memories," Gräff et al. (2014) showed that it is possible to modulate, via drugs, the neuroplasticity of a mouse's brain so that remote traumatic memories (e.g. an event that occurred a month ago) could be treated as efficaciously with exposure-based therapy as recent memories (i.e. day-old events). The mice were first subjected to Pavlovian fear conditioning. At the sound of a tone, they would receive an electrical foot shock. Once the association between the tone and pain was internalized by the mice, they would display a conditioned response at the tone (i.e. they would freeze). They were then treated the day following the establishment of the conditioned response with different behavioural therapies (e.g. exposure therapy) to facilitate fear extinction.

In a different group of mice there was a delay, of thirty days, of exposure to these same behavioural therapies, so that their memories of the association between the tone and pain were *remote* memories rather than recent ones. While the behavioural treatments were successful in extinguishing recent fear memories in the first group of mice, they were unsuccessful in extinguishing the remote fear memories in the other group.

Traumatic events appear to make chemical markers in the genome, and, over time, these markers make it more difficult to respond to behavioural therapy. This is why it is possible to treat the recent but not the remote fearful memories in mice. However, what makes this recent study by Gräff and his colleagues so significant is that the researchers then administered a class of drugs called histone deacetylase inhibitors to the mice. These drugs appear to enhance the neuroplasticity of the brain, extending the period of time needed before traumatic events become ingrained in the genome (and thus more resistant to behavioural therapy). Mice taking this drug, and receiving exposure therapy thirty days after the conditioned response, were able to attenuate the fear response. Such an intervention could be considered a form of MAD because the drug alters the neuroplasticity of the brain so that "recent" memories remain "recent" for a longer period of time, thus extending the window for therapy to be efficacious before fearful events alter the genome.

This study might lead the way to the development of new drugs that permit PTSD in soldiers (and others who suffer traumatic events) to respond better to cognitive behavioural therapy (CBT) by limiting the likelihood that traumatic events will alter the genome in ways that make CBT less efficacious. This could be particularly important for soldiers, as there is often a delay, perhaps of weeks or even months, between the time they experience a traumatic event in a military campaign and when they could begin behavioural therapy.

V The "psychological immune system" and memory modification

The concern that MADs are "unnatural" runs the risk of being predicated upon an overly idealized account of the role memory plays, and should play, in human identity and wellbeing. For example, in *Beyond Therapy*, the President's Council on Bioethics argues that memory-blunting interventions risk "falsifying our perception and understanding of the world" (2003: 228). As such, MADs could threaten the ability of the person having their memory modified from living an *authentic* human life. But is this worry a real danger?

Much depends on what the memory modification does in terms of its impact on our overall wellbeing. I think the President's Council on Bioethics overstates the concern when it claims that the problem is that we risk "falsifying our perception of the world." Such falsifying *is precisely* what the psychological immune system is supposed to do. The key insight worth noting is that some falsifications are unhealthy (e.g. those that allow an alcoholic to remain in denial about their addiction problems), but other falsifications are necessary and beneficial. To see this point, it is worth noting the various ways humans can unconsciously, as well as consciously, already modify their memories.

Having an accurate perception and understanding of the world is not always conducive to our wellbeing. Why? Because life is often full of major disappointments, heartache, disease and death. In *The Wisdom of the Ego*, George Vaillant notes:

> At times we cannot bear reality. At such times our minds play tricks on us. Our minds distort inner and outer reality so that an observer might accuse us of denial, self-deception even dishonesty. But such mental defences creatively re-arrange the sources of our conflict so they become manageable and we may survive. The mind's defences – like the body's immune mechanisms – protect us by providing a variety of illusions to filter pain and to allow self-soothing. (Vaillant 1993: 1)

Daniel Gilbert and his co-authors describe the brain's ability to protect us from gloom as the "psychological immune system." "Ego defense, rationalization, dissonance reduction, motivated reasoning, positive illusions, self-serving attribution, self-deception, self-enhancement, self-affirmation, and self-justification are just some of the terms that psychologists have used to describe the various strategies, mechanisms, tactics, and maneuvers of the psychological immune system" (Gilbert et al. 1998: 619).

Human life always has, and continues to be, rife with pain and suffering. The psychological immune system helps us develop the emotional resilience to cope with adversity and continue to live our lives with optimism and positive emotion. And memory modification is an integral element of our psychological immune system. We have a number of evolved cognitive mechanisms, tactics and strategies that help deny or edit information and facts so that we feel better. These include primitive defence mechanisms such as denial and compartmentalization that can help a person resolve the cognitive dissonance they experience when facing realities that are emotionally painful.

For example, a worker who consistently receives negative job performance reviews might (unconsciously) activate denial to protect his or her ego: "I know I am a great worker, and the only reason I don't get better annual reviews is because I don't schmooze with management like other workers do!" Or a spouse having an extramarital affair might compartmentalize the infidelity by telling themselves: "It's just a bit of fun on the side, nothing serious." These primal defences, while often effective in helping people resolve the pain of cognitive dissonance, are not necessarily helpful in terms of permitting them to live flourishing lives. The temporary pain relief offered by the primitive defence mechanisms can be damaging when they obstruct more healthy ways of working through life's problems to improve the experience of work (e.g. by talking to management about one's annual performance review) and/or relationships (e.g. suggesting marriage counselling). These defence mechanisms impede the growth and reflection of *eudaimonia*.

In "The historical origins of Sigmund Freud's concept of the mechanisms of defense" (chapter 1 in *Ego Mechanisms of Defense*), Vaillant claims that Freud identified five properties of our defence mechanisms:

1 Defenses are a major means of managing instinct and affect.
2 They are unconscious.
3 They are discrete (from one another).
4 Although often the hallmarks of major psychiatric syndromes, defences are dynamic and reversible.
5 They can be adaptive as well as pathological. (Vaillant 1992: 4)

Point 5 is very important to bear in mind for our purposes. Defence mechanisms can be adaptive as well as maladaptive. So any proposed MAD that augments these defence mechanisms, by making them more or less active, could, overall, be potentially beneficial or detrimental to our flourishing.

Denial, for example, while shielding someone from dealing with painful emotions (such as guilt) today can be detrimental to them in the long term, eroding valuable social relationships and even their own health. An alcoholic might self-rationalize their dependency on alcohol by activating their denial defence mechanism ("I'm just a social person, the life of the party!"). But this defence mechanism is maladaptive in this case because the person has become dependent on alcohol to dull painful emotions when they really should be addressing the root cause of this pain (e.g. childhood trauma or abandonment, divorce, unemployment, etc.).

More positive aspects of the psychological immune system, those that can lead to lasting improvements in wellbeing when activated, include humour, optimism and compensation. Laughter, and not taking things too seriously, can help people manage the stresses of life. But when it comes to comedic temperament, having the right amount is crucial. While laughter and a light-hearted attitude to not sweat the small things in life can improve our emotional wellbeing, an adult who treats *all* of life's challenges with dismissive laughter would be a negligent parent, an inefficient worker and an insensitive romantic partner. When it comes to the "comedic virtues," the

Aristotelian doctrine of the mean between extremes is instructive – the right amount of a jovial disposition is the mean between lacking the disposition completely and possessing an excessive amount such that you do not take anything in life seriously.

In her book *Positivity*, the psychologist Barbara Fredrickson (2009) details how optimism and positive emotions help people navigate through the challenges of life. Resilient people are better at transforming negative feelings into positive ones. These persons are not deluded; they also feel frustration and anxiety when facing stressful events. But research has shown that a 3:1 ratio for positive to negative emotions accounts for the difference between individuals that flourish and those that do not (Fredrickson and Losada 2005).

Compensation involves distorting reality in the sense that it downplays shortcomings or failings and emphasizes instead positive attributes or strengths. For example, suppose Betty is asked if Tom, whom she has known as a friend for forty years since they were kids at school, is a good friend. Mary knowns that Tom always shows up late for everything; he has always been this way, even as a child. However, Mary replies "Yes, Tom is a great friend, he tells great stories from the past and is very witty!" Compensation permits Mary to have higher levels of gratitude by focusing on the positives rather than the negatives. Her memory modification enables her to continue to have a positive, healthy relationship with a person with whom she shares a history, despite the fact that he lacks other attributes one might expect or hope a good friend to possess (such as being considerate and punctual).

Cognitive behavioural therapy is a therapeutic intervention that can help with depression and other disorders by encouraging people to change the patterns that lead them to fixate on negative thoughts. And in his book *Redirect*, the social psychologist Timothy Wilson (2011) details how, through story-editing, people can edit their narratives in ways that make their lives more satisfactory. Suppose, for example, Stanley comes home from work one day to find his wife of thirty years has packed all her clothes and left him for good. Stanley was aware they had persistent marital problems for many years, but his wife leaving him was still a shock and he now suffers immense anguish and tries to make sense of the breakdown.

There are many possible narratives Stanley could tell himself about why the marriage ended. For years his ex-wife had emphasized his shortcomings as a husband – that he was untidy, inconsiderate, unromantic, didn't make enough money, etc. One story Stanley could tell himself is "I'm a lousy person and partner." A second possible narrative Stanley could tell himself when making sense of the marriage breakdown is the simple one provided by his best friend Dave. Dave's story is "Your ex-wife is always nagging and no man could ever make her happy!"

Suppose there was some factual accuracy to both these stories. It was true that Stanley wasn't the tidiest of people and that he could have tried to be more romantic towards his wife. But internalizing that version as the *complete* story will lead to chronic low self-esteem, maybe even depression. Such story-editing is maladaptive and corrosive to Stanley's realizing *eudaimonia*. On the flip side, there may also be some truth to the claim that his ex-wife was often overcritical of him (and perhaps other people in her life). If Stanley internalizes the narrative constructed by Dave, this could also be maladaptive as it will prevent him from doing the internal work needed to change the behaviours that contributed to the marriage breakdown (e.g. having better communication with his partner).

The story-editing that Stanley needs to carry out to emerge from his divorce as a healthy, happier person in the long term must strike the right balance between offering some frank assessment of how his behaviour, and his ex-partner's behaviour, led to the relationship ending. And at the same time that story should be compatible with an optimistic outlook that Stanley could, one day, attract and keep a new partner. The right story for Stanley to construct is not necessarily the most *factually accurate* one of the past but, rather, one that helps him grieve and get closure on the past, allows him to make the necessary changes to his behaviour and attitude, and permits him to retain an optimism that he will find fulfilling companionship again in the future.

I have inserted this brief overview of the psychological immune system in this chapter to make clear that humans are *constantly* modifying their memories and editing the story of their lives. An authentic human life is not one of constructing a self-narrative predicated upon a complete, *factually accurate* account of all past events. We are selective, indeed very selective, about what we remember and the weight and meaning we give to positive and negative events. Sometimes this memory alteration is the result of primal defence mechanisms such as denial that protect our ego in ways that can be detrimental, *all-things-considered*, to our wellbeing. But at other times more positive features of this immune system, such as a sense of humour and an optimistic disposition, truly do make our lives go better by eliciting positive emotions and allowing us to live our lives with gratitude and meaning (e.g. "I had to endure fifteen job rejections to become the person I am today, a determined person who will land her dream job any day now!").

When faced with more serious traumatic events – violence, divorce, death of a loved one, etc. – cognitive behavioural therapy or story-editing techniques can be utilized to try to reduce the persistence of negative thoughts and mood, thoughts that might come from an accurate perception of the world and one's life. Because such interventions have been demonstrated to be safe and efficacious, few object to their utilization on the grounds that they are "unnatural" or that the patients undergoing such treatment go on to live "inauthentic" human lives.

Highlighting all of the above points in the previous sections helps, I hope, quell the strong anti-enhancement sentiments that are typically raised against the prospect of developing MADs – namely that they are objectionable because they are "unnatural" or threaten an *authentic* human life. We naturally modify our memories to avoid emotional pain; it's a capacity that helped humans survive in the precarious environments we have lived in throughout our species' evolutionary history.

VI Peace: a cheap and effective way to protect soldiers from the emotional costs of war

When a polity pursues military intervention it puts at risk the lives and physical and mental wellbeing of the men and women serving on the front lines. Asking combatants to risk making the ultimate sacrifice during a military campaign comes with heavy moral responsibilities. There are at least two general moral obligations a virtuous polity must fulfill when requesting (or requiring) soldiers to engage in military intervention. Firstly, there is a moral obligation to try to prevent, as far as is reasonable, putting a soldier's life and health at risk by ensuring they are well equipped with the resources (e.g. armour) needed to successfully complete their mission with the smallest risk of casualties and injury. Secondly, in the event that soldiers sustain physical and/or mental injuries during military operations, medical treatment and support should be provided to them. This duty could entail, for example, providing prosthetic limbs to soldiers injured by roadside bombs and/or cognitive behavioural therapy for conditions such as PTSD.

These two general moral duties could be described as *a duty to prevent harm* to our combatants and *a duty to provide therapy* for those who are injured. Both duties arise from the virtue of benevolence. A country could fail the first duty – the duty to prevent harm – by not providing adequate training and/or equipment to soldiers. Sending soldiers into battle without proper tactical training could place them at a much higher risk of injury and/or death when facing opponents with a much higher degree of skill. Similarly, sending them into battle without adequate equipment (e.g. armour, communications, effective weapons) could increase the probability that they will sustain injury and/or death.

While we send soldiers into the military arena with modernized weapons and armour, they still possess a psychological immune system designed to manage the emotional trauma of our primitive evolutionary past. Is there also a duty to upgrade their emotional resilience if doing so means fewer will suffer the emotional trauma of conditions such as PTSD? So far in this chapter I have tried to undermine the case against enhancing soldiers in this way. Objections that MADs are "unnatural" and might rob soldiers of an authentic human life are less persuasive once we acknowledge that the stressors of modern warfare are also "unnatural" and that modifying memories is a normal part of the psychological immune system.

But the case in favour of MADs will only be compelling provided such interventions actually improve, *all-things-considered*, the quality of life soldiers can live after they are done fighting. A drug that dulled the memories of war by making the primal defence mechanisms of denial hyperactive (e.g. "What war?") would threaten an authentic life in the way the *Beyond Therapy* report noted because it would erode the realization of *eudaimonia*. But a MAD that enhanced the neuroplasticity of the brain so that soldiers who had the remote memories of traumatic events could respond better to CBT could be both very desirable and considered an extension of the duty to prevent harm.

Any presumption in favour of providing MADs to soldiers to reduce the emotional toll inflicted upon them by military conflict must not diminish a prior, and even more pressing moral duty – namely the duty to not place them in the hostile environment of warfare unless doing so promotes a very pressing and substantial moral imperative (e.g. to prevent a grave humanitarian disaster, more war, etc.). However, given we live in a non-ideal world that consists of threats of terrorism, political instability, etc., the prospect of just pursuing peace is not likely to be a feasible and effective solution in the foreseeable future. In the real, non-ideal world MADs might be an effective way of minimizing the extent to which the emotional turmoil of warfare is inflicted upon soldiers when pursuing peaceful solutions are not always viable.

In this chapter I have argued that the objection that MADs are "unnatural" or threaten an authentic human life is not very persuasive. Much of course depends on the specifics of what a MAD actually does. If novel MADs actually help prevent and/or treat PTSD in a manner that improves a soldier's welfare in the long term, then it is hard to see how their use could be unethical provided doing so does not lower the moral threshold for ascertaining when military intervention is justified in the first place. Policy-makers should keep an open mind about the prospects of enhancing the emotional resilience of soldiers via MADs.

VII Modifying behaviour

In earlier chapters we have already noted that genetic heritage has an important influence on human behaviour. From our risks for addiction to our personality and intelligence, the genes we inherit, coupled with the environments in which we are reared and to which we continue to be exposed as adults, help shape our behaviour. This is the complex interplay between nature and nurture. Advances in our understanding of behavioural genetics raise the prospect that, perhaps one day, we might be able to directly alter our nature by editing our genome, or at least purposively altering the expression of particular genes, to reduce our susceptibility to addiction, improve our cognition, or make us more empathetic and moral, etc.

Philosophers have long debated human nature and the moral

motivations of humans. Are people, by nature, altruistic or selfish? The seventeenth-century philosopher Thomas Hobbes argued that humans were primarily self-interested, governed by *felicity* – the insatiable pursuit of power to satisfy our desires. In the "state of nature" – a scenario without government or social cooperation – Hobbes believed life for humans was "nasty, brutish and short." It was only through the creation of government, when people have reasons to believe compliance with a social contract could realize peace, that it was rational for us to behave morally. This stands in contrast to Aristotle's conception of the human being as an inherently "political" animal, pursuing a life of *eudaimonia* (happiness).

While moral and political philosophers might disagree about human nature, what they all did agree about was the importance of *culture* in shaping our behaviour, for better or for worse. Karl Marx, for example, believed that capitalism, an economic system where the vast majority of the population were forced to sell their labour power for wages, exploited and alienated workers and created an antagonistic class system. Jean-Jacques Rousseau argued that the creation of private property corrupted our moral motivations, leading human society down a path where vanity and pride replaced our natural concern for self-preservation (what Rousseau called *amour de soi*). And the eighteenth-century feminist Mary Wollstonecraft argued that the inequality that persisted between men and women was socially created rather than being natural.

Should we aspire to modulate human behaviour via genetic rather than just environmental interventions? Regardless of which account of human nature a normative theorist begins with, there is no denying the fact that humans often fall short of making moral decisions in many aspects of their lives. We might fail to meet *self-regarding* moral duties when we neglect to take care of our physical or mental health because we did not give appropriate weight to long-term rather than short-term (impulsive) goals. Or we might be apathetic about the state of our local or global politics. A virtuous polity would pursue environmental interventions (e.g. public education) to help foster moral and intellectual virtue and combat the risks of moral and intellectual vice. But what about a genetic intervention to improve human nature? If it were feasible to make us more moral by changing our nature, would this be desirable?

Thomas Douglas defines a "moral enhancement" as follows: "A

person morally enhances herself if she alters herself in a way that may reasonably be expected to result in her having morally better future motives, taken in sum, than she would otherwise have had" (Douglas 2008: 229). Defined in this general way, we can see how many different types of interventions could be considered as "moral enhancements," even if they take our nature as a "given." And that is because human nature is not fixed or unitary. Humans can be selfish and violent, but we can also be altruistic and compassionate. The environment in which we are raised during childhood and adolescence (e.g. parental neglect and abuse vs. care and emotional support) influences which elements of our nature are most likely to be activated. And the culture of our society can activate, or constrain, elements of our nature.

Consider, for example, the use of what the social scientist Jon Elster (2000) calls "pre-commitment devices." A pre-commitment involves a conscious effort to bind our future decisions so they are more prudent and/or moral. For example, if a person always tends to make impulsive purchases, causing long-term financial hardship for them and their family, they might decide to rip up their credit cards in an effort to dissuade their future self from the temptation to make more such purchases. The decision to rip up credit cards is a pre-commitment device: it sends a message to your future self that your rational self does not want you to buy anything else you cannot afford. And a democracy with a constitution that protects the basic rights and freedoms of the citizenry *pre-commits* itself to core principles and values that help guard against the "tyranny of the majority."

If our genes influence our behaviour – including our levels of self-control, propensity towards violence, cognitive biases and prosocial behaviour – should we "pre-commit" ourselves to making better moral judgements and individual and collective decision-making by improving our *genetic predisposition* towards immoral behaviour? Prosocial behaviours such as sharing, sympathy and empathy are extremely important for sustaining a polity over time and can help a polity make progress on problems such as poverty, pollution and climate change.

"Cognitive empathy, defined as the ability to recognize what another person is thinking or feeling, and to predict their behaviour based on their mental states, is vital for interpersonal relationships, which in turn is a key contributor of wellbeing" (Warrier et al. 2017:

1). The first large-scale genetic study investigating the genetic architecture of cognitive empathy published its findings in 2017 (ibid.). Researchers recruited customers from 23andMe, a private personal genomics company that, for a fee, analyzes the DNA of a saliva sample. Approximately 89,000 participants of European ancestry completed an online version of the "Reading the mind in the eyes" test. In this test, participants are shown photographs of the eye regions and have to identify the emotion they think the eyes best express. The study identified a genetic locus that is associated with scores on the eyes test in females, "which may be partly due to different genetic architectures in males and females, interacting with postnatal social experience" (ibid.: 7).

Would it be desirable to improve the cognitive, emotional or motivational constraints that impede moral progress? When we consider that question with respect to environmental interventions, we are typically inclined to answer "yes, of course!" Allen Buchanan notes that we have already enhanced our cognitive capacities dramatically – "through literacy, numeracy, science, and electronic, digitized technologies" (2011: 151). Why not pursue further improvements in human nature by directly modifying our biology via genome editing, if this would improve our ability to exercise moral and intellectual vice? One objection, which we have already considered in a general fashion in the introduction, comes in the form of a precautionary principle. That is, if we purposively aspire to change human behaviour by modifying our genes or biology in novel ways, we might create a situation with irreversible, unforeseen grave consequences. But such a risk-averse principle is unsound because it ignores the reality that maintaining the status quo also has harms – harms that are certain (rather than just theoretical) and often non-trivial.

We know that humans are capable of murder, theft, war, genocide, etc. Indeed, some have argued that the moral defects of humans could be amplified as technology progresses further this century (especially the prospect of enhancing human cognition). Ingmar Persson and Julian Savulescu (2008) argue that cognitive enhancement without moral enhancement poses significant dangers to the future of humanity. They summarize their argument like this:

1 It is comparatively easy to cause great harm, much easier than to benefit to the same extent.

2 With the progress of science, which would be speeded up by cognitive enhancement, it becomes increasingly possible for small groups of people, or even single individuals, to cause great harms to millions of people, e.g. by means of nuclear or biological weapons of mass destruction.

3 Even if only a tiny fraction of humanity is immoral enough to want to cause largescale harm by weapons of mass destruction in their possession, there are bound to be some such people in a huge human population, as on Earth, unless humanity is extensively morally enhanced. (Or the human population is drastically reduced, or there is mass genetic screening and selection, though we take it that there is no morally acceptable way of achieving these sufficiently effectively.)

4 A moral enhancement of the magnitude required to ensure that this will not happen is not scientifically possible at present and is not likely to be possible in the near future.

5 Therefore, the progress of science is in one respect for the worse by making likelier the misuse of ever more effective weapons of mass destruction, and this badness is increased if scientific progress is speeded up by cognitive enhancement, until effective means of moral enhancement are found and applied. (Persson and Savulescu 2008: 174)

Improvements in our cognition that improve our capacity for intellectual virtue (or minimize the potential for vice), such as having a sensitivity to details or adaptability of intellect or even intellectual humility, would be morally desirable from the standpoint of the account of virtue ethics I have defended in this book. But it is difficult to speculate on what the potential pros and cons of a genetic moral enhancement would be given how hypothetical and vague such an intervention is (at least at this stage). As the science progresses, it is certainly a topic for more serious reflection and discussion before formulating any concrete prescriptions.

Human happiness, no matter what one takes that term to mean (e.g. positive affect or *eudaimonia*) is influenced by both environment and our genetic makeup. As such, it would be folly to ignore the value of new insights into the role genes play in influencing our behaviour, emotions, motivations, etc. Justice and benevolence prescribe that we pursue environmental interventions, such as imposing criminal sanctions on serious moral wrongdoing and providing decent public education, etc., to try to bring out the best in our nature and minimize the risks of harm that can arise from the worst in our nature. The prospect of being able to edit our genes to change human nature presents us with the introspective question: *What is it to be human?* Would alterations to the human genome make us less than human? And why would such a development be morally desirable or problematic? These are some of the most significant questions for philosophers, and society, to grapple with this century. And, as the field of behavioural genetics advances, the answer to these questions could have great significance in terms of how we decide to use this information to improve our opportunities to live flourishing lives.

Discussion questions

1 Utilitarianism is the view that the morally right action or policy is that which creates the greatest happiness of the greatest number. The genes with which we are born certainly have an important influence on our happiness. Should the regulation of genetic knowledge and technologies be driven by the utilitarian prescription to maximize overall happiness? Why/why not?

2 What do you think about the prospect of developing novel ways to modulate our memory? Would such technologies make us "less human"? What are the potential pros and cons of such technological developments?

3 What is "human nature"? Humans have devised many different cultural interventions to curb our potential for moral vices through interventions such as religion, laws, societal norms, etc. Is there anything *inherently* problematic about aspiring to use genetic interventions, for example genome editing, to alter human behaviour?

Conclusion

Writing in the year 2000, a year before the publication of the draft versions of the human genome (International Human Genome Sequencing Consortium 2001; Venter et al. 2001), four leading bioethicists had the following to say when reflecting upon the potential impact advances in genetic interventions might have on the future of humanity:

> No one knows the limits of our future powers to shape human lives – or when these limits will be reached. Some expect that at most we will be able to reduce the incidence of serious genetic diseases and perhaps ensure that most people are at the higher end of the distribution of normal traits. More people may have long and healthy lives, and perhaps some will have better memory and other intellectual powers. Others foresee not only greater numbers of people functioning at higher levels, but the attainment of levels previously unheard of: lives measured in centuries, people of superhuman intelligence, humans endowed with traits presently undreamt of. One thing, however, is certain: Whatever the limits of our technical abilities turn out to be, coping with these new powers will tax our wisdom to the utmost. (Buchanan et al. 2000: 1)

In this book I have tried to persuade my readership that the virtue ethics tradition is well positioned to help us navigate through the new moral terrain created by advances in our understanding of human biology precisely because it is an ethical theory designed to help us exercise wisdom (or *phronesis*). Unlike moral theories which emphasize the realization of particular principles (e.g. "the greatest happiness of the greatest number," or "liberty upsets patterns"), virtue ethics is a moral tradition that makes *qualities of character* a central focus of moral evaluation. As such, I believe it is a moral theory that can help us think rationally and ethically about the ethical and societal implications of advances in our understanding of genetics and the prospect of genetic interventions.

Should we aspire to "genetically engineer" humans via gene therapy or genome editing? Would a virtuous polity attempt "to improve the biological character of a breed by deliberate methods adopted to that end"? And, if so, how would it do so in ways that help it exercise the moral virtues of benevolence and justice while minimizing the risks of moral and epistemic vice? These and many other questions have been our focus in this book. And the answers or conclusions I have attempted to make and defend are highly provisional. They are meant to help stimulate further reflection and debate rather than settle these contentious issues.

I have employed what I believe is a distinctive moral analysis of moral and epistemic virtue as we navigated through an examination of different aspects of the genetic revolution. The moral virtues of benevolence and justice encourage us to keep an open mind about the ways in which knowledge and innovation can help us prevent harm to individuals and populations, as well as redress inequalities and demonstrate respect for reproductive rights and the health prospects of humans at all stages of the lifespan.

The epistemic virtues of recognizing the salient facts, intellectual humility, adaptability of intellect and the detective's virtues require interdisciplinary engagement and dialogue. These epistemic virtues encourage us to have an accurate understanding of the health challenges facing human populations today and to appreciate the triumphs, but also the limitations, of epidemiology and "negative biology." An adaptive intellect is one that aspires to think outside the box, to find new ways that promote human health and the opportunities for us to enjoy flourishing lives. And yet the virtue of intellectual humility instructs us to have caution and foresight, to understand that interventions in sex selection or age retardation might have unintended consequences, such as altering the sex ratio or intensifying pressures of population density. Thus an ethical regulation of such technological advances must take seriously legitimate societal concerns so that the benefits of such interventions can be realized while minimizing the risks of harmful effects.

This book has defended five provisional moral conclusions:

1 A virtuous polity would determine if any genetic intervention, whether it be gene therapy, genome editing or a drug that activates the expression of specific genes, is morally permissible, indeed perhaps even morally required, by its potential to prevent harm in a reasonably safe and cost-effective manner (e.g. by preventing, delaying or treating morbidity).
2 Virtuous agents would eschew both genetic determinism *and* environmental determinism.
3 A virtuous polity would not necessarily eschew eugenics, where eugenics is understood, as Bertrand Russell defines it, as "the attempt to improve the biological character of a breed by deliberate methods adopted to that end" (1929: 254). In other words, to describe an intervention as "eugenic" does not mean it is unjust. Eugenic aspirations can be morally defensible, even morally obligatory, when they pursue empirically sound and morally justified aims (e.g. promotion of health) through reasonable and morally justified means that treat all persons as free and equal.
4 A virtuous polity would take a *purposeful* approach to determining the scope and limitations of reproductive and parental freedom. Such an approach will give due consideration to the values of autonomy, wellbeing and equality (without ascribing a primacy to any one of these values).
5 A virtuous polity would aspire to promote the *healthy aging* of its population through all possible means (including interventions that extend the lifespan if doing so increased the *healthspan*). Such measures should be pursued in a responsible manner so that considerations of equity, population size, intergenerational justice and environmental impact are also taken seriously.

I believe the topics covered in this book are important issues to address because, when it comes to resolving the societal challenges posed by the genetic revolution, we do not want to repeat the mistakes of the unjust eugenic policies advocated and pursued in the late nineteenth and early twentieth century. Nor do we want to be apathetic, overly suspicious or antagonistic towards innovative interventions that could improve our health and wellbeing. The standard of the virtuous polity is, I believe, a useful and instructive one, even if it is somewhat elusive and ambiguous. At a minimum it can help us gain clarity concerning what might constitute moral and/or epistemic *vice*. But it can also help stimulate reflection and dialogue concerning what the best course of action or attitude or belief might be. Such actions and beliefs are those that a virtuous agent would pursue or hold if they were in our circumstances.

The biology of humans has a long and varied evolutionary history – a history shaped by the hazards of the external world, such as infectious disease, scarcity of food, intergroup conflict, etc. And humans have crafted various forms of *social engineering* to help redress or minimize some of these external risk factors. Public health and preventative medicine, democratic governance, market economies – these are all forms of social engineering that have shaped a culture that, indirectly, influenced the biology of humans. Technological innovations in food production, coupled with a global economy, mean that billions of people in the world today (but not all) have been emancipated from the daily hunger and risks of starvation which would have been typical for many humans living in earlier historical epochs. And yet, in developed countries, the abundance of relatively cheap high-calorie, low-nutrient foods and beverages has increased the incidence of childhood obesity. Social engineering of any kind – whether it be to modify the technologies of food production or political governance or to expedite economic growth – is not necessarily all good or even desirable. A virtuous polity must continually modify, refine and improve the forms of social engineering it employs to improve its knowledge and technology so that individuals and the polity itself can flourish, rather than flounder.

"Genetically engineering" humans, via gene therapy or genome editing or a drug that modulates aging by activating the "longevity genes," is yet one further possible form of social engineering. The critic might ask why we should seriously consider adding genetic intervention into the possible mix of technologies humans pursue. Our response can highlight the prevalence of genetic disorders, from early onset single-gene conditions to more common multi-factorial conditions. The genes we inherit influence not only our health but also our intelligence, behaviour (e.g. parent investment) and happiness and how we age. The genetic revolution might permit humans to intervene intentionally in the genetic lottery of life in a way that improves our life prospects much further than could be realized if we left our biology to the blind and arbitrary process of evolution by natural selection.

By reducing mortality from infectious diseases and developing medical procedures and pharmaceuticals that permit us to manage multi-morbidity in late life, humans can now survive beyond the "biological warranty period" of seven decades. The aging of human populations is a very new and novel phenomenon and one that demonstrates how important the epistemic virtue of adaptability of intellect is if we hope to improve the health prospects of an aging world. New knowledge about our genes might prove to be foundational in developing the health innovations needed to realize greater equality, health and economic prosperity for all of the world's diverse populations.

Notes

Introduction

1 "Constitution of Who: principles," www.who.int/about/mission/en/.
2 "Gene therapy clinical trials worldwide," www.abedia.com/wiley/conti nents.php.
3 Doubts about this experiment have been raised. See Egli et al. (2017).
4 See, for example, Banerjee (2007).
5 See Glenn et al. (2011).
6 In "Evolutionary perspectives on the obesity epidemic: adaptive, maladaptive, and neutral viewpoints," John Speakman considers this evolutionary explanation for the passing on of "thrifty genes" in detail and finds it untenable. Speakman contends that, rather than being adaptive, obesity is either a maladaptive by-product of positive selection on some other trait or that most mutations in the genes that predispose us to obesity are neutral.
7 Indeed genetic engineering is, I would contend, simply a specific type of social engineering. It utilizes specific kinds of knowledge (e.g. about genetics and evolutionary biology) to develop technologies that could, with the right cultural circumstances (e.g. medical expertise, resources, etc.), permit us to genetically engineer humans through gene therapy or genome editing.
8 I borrow the phrases "Luddites" and "Technophiles" from Gordon Graham (1999) to characterize these two extreme positions.

Chapter 1 Eugenics: Inherently Immoral?

1 "Folic acid: recommendations," www.cdc.gov/ncbddd/folicacid/recom mendations.html.

2 These considerations are important to keep in mind once we turn to consider the prospect of genetic intervention in the chapters to come.
3 "Over 70 years of community water fluoridation," www.cdc.gov/fluori dation/basics/70-years.htm.
4 Ibid.
5 See Dugdale (1910).

Chapter 2 The Genetic Revolution

1 "The cost of sequencing a human genome," www.genome.gov/sequenc ingcosts/.
2 Ibid.
3 A study of PGD use on mice, for example, found that the nervous system may be sensitive to blastomere biopsy procedures, and that this might result in an increased relative risk of neurodegenerative disorders in the subsequent offspring (Yu et al. 2009).
4 There is much debate among scholars as to the extent to which the passage can be attributed to the view of Protagoras or to Plato himself.
5 See "Gene therapy clinical trials worldwide," www.abedia.com/wiley/ continents.php for data on gene therapy clinical trials worldwide.
6 US Food and Drug Administration, "Step 3: Clinical research," www. fda.gov/ForPatients/Approvals/Drugs/ucm405622.htm#Clinical_Resear ch_Phase_Studies.
7 "Why mouse matters," www.genome.gov/10001345/importance-of-mouse-genome/.
8 The discussion that follows draws upon the insights argued for in this earlier article.
9 Luck egalitarians maintain that inequalities in the advantages that people enjoy are just if they derive from the choices people have voluntarily made; however, inequalities deriving from unchosen features of people's circumstances are unjust.
10 The libertarian Robert Nozick (1974) argues that redistributive taxation is unjust because it violates self-ownership.
11 Pew Research Center, "Public's views on human evolution," www.pew forum.org/2013/12/30/publics-views-on-human-evolution/.
12 "What is genetic discrimination?," https://ghr.nlm.nih.gov/primer/ testing/discrimination.

Chapter 3 Disease

1 *The Cancer Genome Atlas*, https://cancergenome.nih.gov/.
2 World Health Organization, "Cancer," www.who.int/mediacentre/fact sheets/fs297/en/.

3 *Progress on Sanitation and Drinking Water: 2015 Update and MDG Assessment*, www.unicef.org/publications/index_82419.html.
4 "Sickle cell disease: data & statistics," www.cdc.gov/ncbddd/sicklecell/data.html.
5 "The impact of chronic disease in high income countries," www.who.int/chp/chronic_disease_report/media/hi_income.pdf.
6 "The impact of chronic disease in lower middle income countries," www.who.int/chp/chronic_disease_report/media/lower_middle.pdf.
7 See Kirkwood (1977) and Kirkwood and Holliday (1979).
8 The concept of a "biological warranty period" is developed by Bruce Carnes et al. (2003).
9 See Colin Farrelly (2012a and 2012b). The next few paragraphs are drawn from both journal articles.

Chapter 4 Epigenetics

1 The original citation is from Goldberg et al. (2007).
2 "Depression," www.who.int/mediacentre/factsheets/fs369/en/.
3 National Institute on Drug Abuse, "Trends & statistics," www.drugabuse.gov/related-topics/trends-statistics.

Chapter 5 Reproductive Freedom

1 F. Newport and J. Wilke, "Desire for children still norm in U.S.," www.gallup.com/poll/164618/desire-children-norm.aspx.
2 See Buchanan et al. (2000) for a more detailed discussion of these concerns.
3 As Shakespeare (2006: 87) notes, prenatal screening could never significantly reduce the incidence of disability. "Approximately 2 per cent of all births are affected by congenital abnormality, whereas disabled people compromise 10–20 per cent of the population, suggesting that genetic screening could never seriously reduce the incidence of disability."
4 See de Melo-Martin (2004: 75–8) for an elaboration of this line of objection.

Chapter 6 Aging Research and Longevity

1 In their extensive cross-cultural examination of longevity in hunter-gatherers, Gurven and Kaplan (2007: 322) conclude that "human bodies are designed to function well for about seven decades in the environment in which our species evolved."
2 Boston University's New England Centenarian Study, at www.bumc.bu.edu/centenarian/overview.

3 See, for example, De Grey (2005).
4 See Brewer et al. 1981; Brewer and Lui 1984; Schmidt and Boland 1986.
5 D. J. Jeffries, "Reply to Colin Farrelly," www.bmj.com/content/337/bmj. a414/rapid-responses.
6 A *zero-sum game* is a situation where, if one person's distribution or situation is improved, it results in another person having less. So, in two-person zero-sum games, only one person can win and the other must be the loser.
7 The relevance of this point for defending aging research was initially developed and argued for in Farrelly (2008).
8 I examine the implications of aging research from Daniels's account of just healthcare more fully in Farrelly (2010).

Chapter 7 Happiness, Memory and Behaviour

1 www.parl.gc.ca/Content/LOP/ResearchPublications/2011-97-e.htm #a2.
2 See Caplan (2005).
3 Perhaps 300,000 years. See Hublin et al. (2017).

References

Agar, N. (2004) *Liberal Eugenics: In Defence of Human Enhancement*. Oxford: Blackwell.

Aidoo, M., et al. (2002) "Protective effects of the sickle cell gene against malaria morbidity and mortality," *The Lancet* 359(9314): 1311–12.

Allhoff, F. (2005) "Germ-line genetic enhancement and Rawlsian primary goods," *Kennedy Institute of Ethics Journal* 15(1): 39–56.

Andersen, S., et al. (2012) "Health span approximates life span among many supercentenarians: compression of morbidity at the approximate limit of life span," *Journals of Gerontology: Series A, Medical Sciences* 67A(4): 395–405.

Anderson, G. F., and Hussey, P. S. (2000) "Population aging: comparison among industrialized countries," *Health Affairs* 19(3): 191–203.

Arias, E. (2002) "United States life tables, 2000," *National Vital Statistics Reports* 51(3): 1–38.

Atzmon, G., et al. (2004) "Clinical phenotype of families with longevity," *Journal of the American Geriatrics Society* 52(2)L 274–77.

Auroux, M. R., et al. (1989) "Paternal age and mental functions of progeny in man," *Human Reproduction* 4(7): 794–7.

Auroux, M. R., et al. (2009) "Progeny's mental aptitudes in man: relationship with parental age at conception and with some environmental factors," *Comptes Rendus Biologies* 332(7): 603–12.

Banerjee, A. V. (2007) *Making Aid Work*. Cambridge, MA: MIT Press.

Bauer, J., et al. (2008) "Narrative identity and eudaimonic well-being," *Journal of Happiness Studies* 9(1): 81–104.

Benyamin, B., et al. (2014) "Childhood intelligence is heritable, highly polygenic and associated with *FNBP1L*," *Molecular Psychiatry* 19(2): 253–8.

Bevilacqua, L., and Goldman, D. (2009) "Genes and addictions," *Clinical Pharmacology & Therapeutics* 85(4): 359–61.

Bok, D. (2010) *The Politics of Happiness*. Princeton, NJ: Princeton University Press.

Bombay, A., et al. (2014) "The intergenerational effects of Indian residential schools: implications for the concept of historical trauma," *Transcultural Psychiatry* 51(3): 320–38.

Boonin, D. (2014) *The Non-Identity Problem and the Ethics of Future People*. Oxford: Oxford University Press.

Bowles, S. (2009) "Did warfare among ancestral hunter-gatherers affect the evolution of human social behaviors?" *Science* 324(5932): 1293–8.

Brewer, M. B., and Lui, L. (1984) "Categorization of the elderly by the elderly," *Personality and Social Psychology Bulletin* 10(4): 585–95.

Brewer, M. B., et al. (1981) "Perceptions of the elderly: stereotypes as prototypes," *Journal of Personality and Social Psychology* 41(4): 656–70.

Bribiescas, R. (2006) *Men: Evolutionary and Life History*. Cambridge, MA: Harvard University Press.

Brock, D. (2005) "Shaping future children: parental rights and societal interests," *Journal of Political Philosophy* 13(4): 377–98.

Brown, J. (2001) "Genetic manipulation in humans as a matter of Rawlsian justice," *Social Theory and Practice* 27(1): 83–110.

Buchanan, A. (1995) "Equal opportunity and genetic intervention," *Social Philosophy and Policy* 12(2): 105–35.

Buchanan, A. (2011) *Beyond Humanity*. Oxford: Oxford University Press.

Buchanan, A., Brock, D. W., Daniels, N., and Wikler, D. (2000) *From Chance to Choice: Genetics and Justice*. Cambridge: Cambridge University Press.

Butler, R. N. (1969) "Age-ism: another form of bigotry," *The Gerontologist* 9(4): 243–6.

Butler, R. N. et al. (2008) "New model of health promotion and disease prevention for the 21st century," *British Medical Journal* 337(7662): 149–50.

Callaghan, D. (1987) *Setting Limits: Medical Costs in an Aging Society*. Washington, DC: Georgetown University Press.

Canadian Cancer Society's Advisory Committee on Cancer Statistics (2017) *Canadian Cancer Statistics 2017*. Toronto: Canadian Cancer Society.

Caplan, A. (2005) "Death as an unnatural process," *EMBO Reports* 6(Suppl. 1): S72–S75.

Carnes, B. (2007) "Senescence as viewed through the lens of comparative biology," *Annals of the New York Academy of Sciences* 1114: 14–22.

Carnes, B., et al. (2003) "Biological evidence for limits to the duration of life," *Biogerontology* 4(1): 31–45.

Caspi, A., et al. (2003) "Influence of life stress on depression: moderation by a polymorphism in the 5-HTT gene," *Science* 301(5631): 386–9.

Champagne, F., and Curley, C. (2012) "Genetics and epigenetics of parental care," in Royle, N. J., Smiseth, P. T., and Kölliker, M. (eds),

The Evolution of Parental Care. Oxford: Oxford University Press, pp. 304–26.

Childs, B. (1999) *Genetic Medicine: A Logic of Disease.* Baltimore: Johns Hopkins University Press.

Choquet, H., and Meyre, D. (2011) "Genetics of obesity: what have we learned?" *Current Genomics* 12(3): 169–79.

Cook-Deegan, R. M., and McCormack, S. J. (2001) "Intellectual property: patents, secrecy, and DNA," *Science*, 293(5528): 217.

Cousino, M., and Hazen, R. (2013) "Parenting stress among caregivers of children with chronic illness: a systematic review," *Journal of Pediatric Psychology* 38(8): 809–28.

Couzin-Frankel, J. (2011) "A pitched battle over life span," *Science* 333(6042): 549–50.

Craig, I., and Halton, W. (2009) "Genetics of human aggressive behavior," *Human Genetics* 126(1): 101–13.

Craig, J. (1994) "Replacement level fertility and future population growth," *Population Trends* 78: 20–2.

Crespi, B. (2000) "Evolution of maladaption," *Heredity* 84: 623–9.

Curley, J., et al. (2011) "Epigenetics and the origins of paternal effects," *Hormones and Behavior* 59(3): 306–14.

Cyranoski, D. (2016) "CRISPR gene-editing tested in a person for the first time," *Nature* 539(7630): 479.

Daniels, N. (1985) *Just Health Care.* Cambridge: Cambridge University Press.

Daniels, N. (2000) "Normal functioning and the treatment–enhancement distinction," *Cambridge Quarterly of Healthcare Ethics* 9: 309–22.

Daniels, N. (2008) *Just Health.* Cambridge: Cambridge University Press.

Deci, E., and Ryan, R. (2008) "Hedonia, eudaimonia, and well-being: an introduction," *Journal of Happiness Studies* 9(1): 1–11.

De Grey, A. (2005) "Life extension, human rights, and the rational refinement of repugnance," *Journal of Medical Ethics* 31(11): 659–63.

de Melo-Martin, I. (2004) "On our obligation to select the best children: a reply to Savulescu," *Bioethics* 18(1): 72–83.

DiMasi, J., et al. (2016) "Innovation in the pharmaceutical industry: new estimates of R&D costs," *Journal of Health Economics* 47(May): 20–33.

Douglas, T. (2008) "Moral enhancement," *Journal of Applied Philosophy* 25(3): 228–45.

Dugdale, R. (1910) *The Jukes: A Study in Crime, Pauperism, Disease, and Heredity.* New York: G. P. Putnam's Sons.

Dunn, E., et al. (2008) "Spending money on others promotes happiness," *Science* 319(5870): 1687–8.

Dupras, C., et al. (2014) "Epigenetics and the environment in bioethics," *Bioethics* 28(3): 327–34.

Egli, D., et al. (2017) "Inter-homologue repair in fertilized human eggs?" *BioRxiv* preprint, doi: http://dx.doi.org/10.1101/181255.

Elster, J. (2000) *Ulysses Unbound: Studies in Rationality, Precommitment, and Constraints.* Cambridge: Cambridge University Press.

Evert, J., et al. (2007) "Morbidity profiles of centenarians: survivors, delayers, and escapers," *Journals of Gerontology: Series A, Biological Sciences and Medical Sciences* 58(3): M232–M237.

Farrelly, C. (2004) "Genes and equality," *Journal of Medical Ethics* 30(4): 587–92.

Farrelly, C. (2008) "Aging research: priorities and aggregation," *Public Health Ethics* 1(3): 258–67.

Farrelly, C. (2010) "Equality and the duty to retard human aging," *Bioethics* 24(8): 384–94.

Farrelly, C. (2012a) "'Positive biology' as a new paradigm for the medical sciences," *EMBO Reports* 13(2): 186–8.

Farrelly, C. (2012b) "Why the NIH should create an institute of positive biology," *Journal of the Royal Society of Medicine* 105: 412–15.

Farrelly, C. (2016) *Biologically Modified Justice.* Cambridge: Cambridge University Press.

Fenner, F., et al. (1988) *Smallpox and its Eradication.* Geneva: World Health Organization.

Finer, L., and Zolna, M. (2016) "Declines in unintended pregnancy in the United States, 2008–2011," *New England Journal of Medicine* 374: 843–52.

Flory, J., and Kitcher, P. (2004) "Global health and the scientific research agenda," *Philosophy & Public Affairs* 4(32): 36–65.

Fowke, K., et al. (1996) "Resistance to HIV-1 infection among persistently seronegative prostitutes in Nairobi, Kenya," *The Lancet* 348(9038): 1347–51.

Fowler, J., and Dawes, C. (2008) "Two genes predict voter turnout," *Journal of Politics* 70(3): 579–94.

Franco, O., et al. (2005) "Effects of physical activity on life expectancy with cardiovascular disease," *Archives of Internal Medicine* 165(20): 2355–60.

Frans, E. M., et al. (2008) "Advancing paternal age and bipolar disorder," *Archives of General Psychiatry* 65(9): 1034–40.

Fredrickson, B. (1998) "What good are positive emotions?" *Review of General Psychology* 2(3): 300–19.

Fredrickson, B. (2009) *Positivity.* New York: Three Rivers Press.

Fredrickson, B., and Losada, M. (2005) "Positive affect and the complex dynamics of human flourishing," *American Psychologist* 60(7): 678–86.

Fries, J. (2005) "The compression of morbidity," *Milbank Quarterly* 83(4): 801–23.

George, R. (1993) *Making Men Moral.* Oxford: Oxford University Press.

Gilbert, D., and Wilson, T. (2005) "Prospection: experiencing the future," *Science* 317(5843): 1351–4.

Gilbert, D., et al. (1998) "Immune neglect: a source of durability bias in affective forecasting," *Journal of Personality and Social Psychology* 75(3): 617–38.

Glannon, W. (2001) *Genes and Future People*. Boulder, CO: Westview Press.

Glenn, A., et al. (2011) "Evolutionary theory and psychopathy," *Aggression and Violent Behavior* 16(5): 371–80.

Goldberg, A. D., et al. (2007) "Epigenetics: a landscape takes shape," *Cell* 128(4): 635–8.

Goldman, D., et al. (2005) "The genetics of addictions: uncovering the genes," *Nature Reviews Genetics* 6(7): 521–32.

Gonzalez, E., et al. (2005) "The influence of CCL3L1 gene-containing segmental duplications on HIV-1/AIDS susceptibility," *Science* 307(5714): 1434–40.

Gottfredson, L. (1997) "Mainstream science on intelligence: an editorial with 52 signatories, history, and bibliography," *Intelligence* 24(1): 13–23.

Gräff, J., et al. (2014) "Epigenetic priming of memory updating during reconsolidation to attenuate remote fear memories," *Cell* 156(1–2): 261–76.

Graham, G. (1999) *The Internet: A Philosophical Inquiry*. London: Routledge.

Guégan, J. F., et al. (2008) "Global spatial patterns of infectious disease," in Stearns, S., and Koella, J. C. (eds), *Evolution in Health and Disease*. Oxford: Oxford University Press, pp. 19–30.

Gurven, M., and Kaplan, H. (2007) "Longevity among hunter-gatherers: a cross-cultural examination," *Population and Development Review* 33(2): 321–65.

Guthrie, W. K. C. (1956) *Plato: Protagoras and Meno*. Harmondsworth: Penguin.

Gutmann, A., and Thompson, D. (2004) *Why Deliberative Democracy?* Princeton, NJ: Princeton University Press.

Habermas, J. (2003) *The Future of Human Nature*. Cambridge: Polity.

Hamilton, J., and Mestler, G. (1969) "Mortality and survival: comparison of eunuchs with intact men and women in a mentally retarded population," *Journal of Gerontology* 24(4): 395–411.

Hampton, J. (1984) "The moral education theory of punishment," *Philosophy and Public Affairs* 13(3): 208–38.

Hanser, M. (1990) "Harming future people," *Philosophy and Public Affairs* 19(1): 47–70.

Harris, J. (1985) *The Value of Life*. London: Routledge & Kegan Paul.

Harris, J. (2007) *Enhancing Evolution: The Ethical Case for Making Better People*. Princeton, NJ: Princeton University Press.

Harrison, D., et al. (2009) "Rapamycin fed late in life extends lifespan in genetically heterogeneous mice," *Nature* 460: 392–5.

Hayflick, L. (2000) "The future of ageing," *Nature* 408(6809): 267–9.

He, F., et al. (2006) "Consequences of paternal cocaine exposure in mice," *Neurotoxicology and Teratology* 28(2): 198–209.

Heller, M., and Eisenberg, R. (1998) "Can patents deter innovation? The anticommons in biomedical research," *Science*, 280: 698–701.

Hesketh, T., and Xing, Z. W. (2006) "Abnormal sex ratios in human populations: causes and consequences," *Proceedings of the National Academy of Sciences* 103(36): 13271–5.

Heyd, D. (1992) *Genethics: Moral Issues in the Creation of People*. Berkeley: University of California Press.

Holloszy, J., and Fontana, L. (2007) "Caloric restriction in humans," *Experimental Gerontology* 42(8): 709–12.

Holmes, S., and Sunstein, C. (1999) *The Costs of Rights: Why Liberty Depends on Taxes*. New York: W. W. Norton.

Hublin, J. J., et al. (2017) "New fossils from Jebel Irhoud, Morocco and the pan-African origin of *Homo sapiens*," *Nature* 546: 289–92.

International Human Genome Sequencing Consortium (2001) "Initial sequencing and analysis of the human genome," *Nature* 409: 860–921.

Kaati, G., et al. (2002) "Cardiovascular and diabetes mortality determined by nutrition during parents' and grandparents' slow growth period," *European Journal of Human Genetics* 10(11): 682–8.

Kaeberlein, M., et al. (2015) "Healthy aging: the ultimate preventative medicine," *Science* 350(6265): 1191–3.

Kahneman, D., and Deaton, A. (2010) "High income improves evaluation of life but not emotional well-being," *Proceedings of the National Academy of Sciences of the United States of America* 107(38): 16489–93.

Kahneman, D., and Riis, J. (2005) "Living and thinking about it: two perspectives on life," in Huppert, F. A., Baylis, N., and Keverne, B. (eds), *The Science of Well-Being*. Oxford: Oxford University Press, pp. 285–301.

Kahneman, D., et al. (2006) "Would you be happier if you were richer? A focusing illusion," *Science* 312(5782): 1908–10.

Karlsson, E., et al. (2014) "Natural selection and infectious disease in human populations," *Nature Reviews Genetics* 15(6): 379–93.

Kennedy, B., et al. (2014) "Geroscience: linking aging to chronic disease," *Cell* 159(4): 709–13.

Kirkwood, T. (1977) "Evolution of aging," *Nature* 270: 301–4.

Kirkwood, T., and Austad, S. (2000) "Why do we age?" *Nature* 408(9): 233–8.

Kirkwood, T., and Holliday, R. (1979) "The evolution of ageing and longevity," *Proceedings of the Royal Society of London: Biology* 205(1161): 531–46.

Koplow, D. (2003) *Smallpox: The Fight to Eradicate a Global Scourge.* Berkeley: University of California Press.

Kourany, J. (2014) "Human enhancement: making the debate more productive," *Erkenntnis* 79(Suppl. 5) 981–98.

Kraut, R. (2002) *Aristotle: Political Philosophy.* Oxford: Oxford University Press.

Kuliev, A., and Verlinsky, Y. (2005) "Preimplantation diagnosis: a realistic option for assisted reproduction and genetic practice," *Current Opinion in Obstetrics and Gynecology* 17(2): 179–83.

Kyung-Jin, M., et al. (2012) "The lifespan of Korean eunuchs," *Current Biology* 22(18): R792–R793.

LeBlanc, S. (2014) "Forager warfare and our evolutionary past," in Allen, M. W., and Jones, T. L. (eds), *Violence and Warfare among Hunter-Gatherers.* Walnut Creek, CA: Left Coast Press, pp. 26–46.

Lerner, G. (1986) *The Creation of Patriarchy.* New York: Oxford University Press.

Levy, S., et al. (2007) "The diploid genome sequence of an individual human," *PLoS Biology* 5(10): 2113–44.

Liao, M., and Sandberg, A. (2008) "The normativity of memory modification," *Neuroethics* 1(2): 85–99.

Liebaers, I., et al. (2010) "Report on a consecutive series of 581 children born after blastomere biopsy for preimplantation genetic diagnosis," *Human Reproduction* 25(1): 275–82.

Lundstrom, S., et al. (2010) "Trajectories leading to autism spectrum disorders are affected by paternal age: findings from two nationally representative twin studies," *Journal of Child Psychology and Psychiatry* 51(7): 850–6.

Lykken, D., and Tellegen, A. (1996) "Happiness is a stochastic phenomenon," *Psychological Science* 7(3): 186–9.

Ma, H., et al. (2017) "Correction of a pathogenic gene mutation in human embryos," *Nature* 548: 413–19.

Malaspina, D., et al. (2001) "Advancing paternal age and the risk of schizophrenia," *Archives of General Psychiatry* 58(4): 361–7.

Malaspina, D., et al. (2005) "Paternal age and intelligence: implications for age-related genomic changes in male germ cells," *Psychiatric Genetics* 15(2): 117–25.

McDonald, M., et al. (2012) "Evolution and the psychology of intergroup conflict: the male warrior hypothesis," *Philosophical Transactions of the Royal Society of London B: Biological Sciences* 367(1589): 670–9.

McDougall, R. (2007) "Parental virtue: a new way of thinking about the morality of reproductive actions," *Bioethics* 21(4): 181–90.

McIntosh, G. C., et al. (1995) "Paternal age and the risk of birth defects in offspring," *Epidemiology* 6(3): 282–8.

Medawar, P. (1952) *An Unsolved Problem of Biology*. London: Lewis.

Mehl, M., et al. (2010) "Eavesdropping on happiness: well-being is related to having less small talk and more substantive conversations," *Psychological Science* 21(4): 539–41.

Mill, J. S. ([1859] 1956) *On Liberty*, ed. C. V. Shields. Indianapolis: Bobbs-Merrill.

Miller, R. (2002) "Extending life: scientific prospects and political obstacles," *Milbank Quarterly* 80(1): 155–74.

Moore, L., et al. (2013) "DNA methylation and its basic function," *Neuropsychopharmacology* 38(1): 23–38.

Moore, M. (1987) "The moral worth of retributivism," in Shoeman, F. (ed.), *Character, Responsibility and the Emotions*. New York: Cambridge University Press, pp. 179–219.

Mustanski, B. S., et al. (2005) "A genomewide scan of male sexual orientation," *Human Genetics* 116(4): 272–8.

NASEM (National Academies of Sciences, Engineering, and Medicine) (2016) *Genetically Engineered Crops: Experiences and Prospects*. Washington, DC: National Academies Press.

NASEM (National Academies of Sciences, Engineering, and Medicine) (2017) *Human Genome Editing: Science, Ethics, and Governance*. Washington, DC: National Academies Press.

Nelson, S. K., et al. (2013) "In defense of parenthood: children are associated with more joy than misery," *Psychological Science* 24(1): 3–10.

Nozick, R. (1974) *Anarchy, State and Utopia*. New York: Basic Books.

Okbay A., et al. (2016) "Genetic variants associated with subjective well-being, depressive symptoms, and neuroticism identified through genome-wide analyses," *Nature Genetics* 48(6): 624–33.

Olshansky, S. J., et al. (1990) "In search of Methuselah: estimating the upper limits to human longevity," *Science* 250(4981): 634–40.

Olshansky, S. J., et al. (1993) "The aging of the human species," *Scientific American* 268(4): 46–52.

Olshansky, S. J., et al. (1998) "Confronting the boundaries of human longevity," *American Scientist* 86(1): 52–61.

Olshansky, S. J., et al. (2005) "A potential decline in life expectancy in the United States in the 21st century," *New England Journal of Medicine* 352: 1138–45.

Olshansky, S. J., et al. (2006) "In pursuit of the longevity dividend," *The Scientist* 20(March): 28–36.

Orzack, S., et al. (2015) "The human sex ratio from conception to birth," *Proceedings of the National Academy of Sciences* 112(16): E2102–E2111.

Overall, C. (2012) *Why Have Children?* Cambridge, MA: MIT Press.

Palmore, E. (1999) *Ageism: Negative and Positive*. New York: Springer.

Pappas, C., et al. (2008) "Single gene reassortants identify a critical role for

PB1, HA, and NA in the high virulence of the 1918 pandemic influenza virus," *Proceedings of the National Academy of Sciences* 105(8): 3064–9.

Parfit, D. (1984) *Reasons and Persons*. Oxford: Oxford University Press.

Pchelin, P., and Howell, R. (2014) "The hidden cost of value-seeking: people do not accurately forecast the economic benefits of experiential purchases," *Journal of Positive Psychology* 9(4): 322–34.

Perls, T. (1997) "Centenarians prove the compression of morbidity hypothesis, but what about the rest of us who are genetically less fortunate?" *Medical Hypotheses* 49(5): 405–7.

Perls, T., et al. (1998) "Siblings of centenarians live longer," *The Lancet* 351(9115): 1560.

Persson, I., and Savulescu, J. (2008) "The perils of cognitive enhancement and the urgent imperative to enhance the moral character of humanity," *Journal of Applied Philosophy* 25(3): 162–77.

Pérusse, D., et al. (1994) "Human parental behavior: evidence for genetic influence and potential implication for gene-culture transmission," *Behavior Genetics* 24(4): 327–35.

Plato (1997) *Republic*, in *Complete Works*, ed. J. Cooper. Indianapolis: Hackett.

Plomin, R., and Deary, I. J. (2015) "Genetics and intelligence differences: five special findings," *Molecular Psychiatry* 20(1): 98–108.

Pollmann-Schult, M. (2014) "Parenthood and life satisfaction: why don't children make people happy?" *Journal of Marriage and Family* 76(2): 319–36.

Popenoe, P. (1935) "Education and eugenics," *Journal of Educational Sociology* 8(8): 451–8.

Powell, R., and Buchanan, A. (2011) "Breaking evolution's chains: the prospect of deliberate genetic modification in humans," *Journal of Medicine and Philosophy* 36(1): 6–27.

President's Council on Bioethics (2003) *Beyond Therapy: Biotechnology and the Pursuit of Happiness*, https://bioethicsarchive.georgetown.edu/pcbe/reports/beyondtherapy/index.html.

Rawls, J. (1971) *A Theory of Justice*. Cambridge, MA: Belknap Press.

Rawls, J. (1999) "Two concepts of rules," in *Collected Papers*, ed. S. Freeman. Cambridge, MA: Harvard University Press, pp. 20–46.

Reichenberg, A., et al. (2006) "Advancing paternal age and autism," *Archives of General Psychiatry* 63: 1026–32.

Resnik, D. (1997) "Genetic engineering and social justice: a Rawlsian approach," *Social Theory and Practice* 23(3): 427–48.

Resnick, D. (2000) "The moral significance of the therapy-enhancement distinction in human genetics," *Cambridge Quarterly of Healthcare Ethics* 9: 365–77.

Resnick, D. (2012) "Ethical virtues in scientific research," *Accountability in Research* 19(6): 329–43.

Roberts, M., and Wasserman, D. (eds) (2009) *Harming Future Persons: Ethics, Genetics and the Nonidentity Problem.* Dordrecht: Springer.

Rose, M. (2005) *The Long Tomorrow.* New York: Oxford University Press.

Russell, B. (1929) "Eugenics," in *Marriage and Morals.* New York: Liveright, pp. 254–73.

Sandel, M. (2007) *The Case against Perfection: Ethics in the Age of Genetic Engineering.* Cambridge, MA: Harvard University Press.

Savulescu, J. (2001) "Procreative beneficence: why we should select the best children," *Bioethics* 15(5–6): 413–26.

Savulescu, J., and Kahane, G. (2009) "The moral obligation to create children with the best chance of the best life," *Bioethics* 23(5): 274–90.

Scanlon, T. (1998) *What We Owe to Each Other.* Cambridge, MA: Harvard University Press.

Schmidt, D. F., and Boland, S. M. (1986) "Structure of perceptions of older adults: evidence for multiple stereotypes," *Psychology and Aging* 1(3): 255–60.

Schüklenk, U., et al. (2011) "End-of-life decision-making in Canada: the report by the Royal Society of Canada Expert Panel on End-of-Life Decision-Making," *Bioethics* 25(S1): 1–73.

Seligman, M. (2002) *Authentic Happiness.* New York: Free Press.

Shakeshaft, N., et al. (2015) "Thinking positively: the genetics of high intelligence," *Intelligence* 48: 123–32.

Shakespeare, T. (2006) *Disability Rights and Wrongs.* Abingdon: Routledge.

Sibbald, B. (2001) "Death but one unintended consequence of gene-therapy trial," *Canadian Medical Association Journal* 164(11): 1612.

Simonsen, L., et al. (1998) "Pandemic versus epidemic influenza mortality: a pattern of changing age distribution," *Journal of Infectious Diseases* 178(1): 53–60.

Singer, P. (1972) "Famine, affluence and morality," *Philosophy and Public Affairs* 1(3): 229–43.

Smietana, K., et al. (2016) "Trends in clinical success rates," *Nature Reviews Drug Discovery* 15: 379–80.

Smith, M., and Harper, D. (1988) "The evolution of aggression: can selection generate variability?" *Philosophical Transactions of the Royal Society of London, Series B: Biological Science* 219(1196): 557–70.

Smuts, B. (1995) "The evolutionary origins of patriarchy," *Human Nature* 6(1): 1–32.

Speakman, J. (2013) "Evolutionary perspectives on the obesity epidemic: adaptive, maladaptive, and neutral viewpoints," *Annual Review of Nutrition* 33: 289–317.

Stearns, S., et al. (2008) "Introducing evolutionary thinking for medicine," in Stearns, S., and Koella, J. (eds), *Evolution in Health and Disease.* 2nd edn, Oxford: Oxford University Press, pp. 3–16.

Stephenson, A., et al. (2015) "A contemporary survival analysis of individuals with cystic fibrosis: a cohort study," *European Respiratory Journal*
45(3): 670–9.

Sunstein, C. (2003) "Beyond the precautionary principle," *University of
Pennsylvania Law Review* 151: 1003–58.

Swartz, J. R., et al. (2017) "An epigenetic mechanism links socioeconomic
status to changes in depression-related brain function in high-risk adolescents," *Molecular Psychiatry* 22(2): 209–14.

Tabatabaie, V., et al. (2011) "Exceptional longevity is associated with
decreased reproduction," *Aging* 3(12): 1202–5.

Tanser, F., et al. (2003) "Potential effect of climate change on malaria
transmission in Africa," *The Lancet* 362(9398): 1792–8.

Taylor, L., et al. (2001) "Risk factors for human disease emergence,"
Philosophical Transactions of the Royal Society of London B 356: 983–9.

Temkin, L. (1987) "Transitivity and the mere addition paradox," *Philosophy
and Public Affairs* 16(2): 138–87.

Tiihonen, J., et al. (2015) "Genetic background of extreme violent behavior," *Molecular Psychiatry* 20(6): 786–92.

Turner, D. C., et al. (2003) "Cognitive enhancing effects of modafinil in
healthy volunteers," *Psychopharmacology* 165: 260–9.

United Nations, Department of Economic and Social Affairs, Population
Division (2011) *World Population Prospects: The 2010 Revision, Highlights
and Advance Tables*. New York: United Nations.

Vaillant, G. (1992) *Ego Mechanisms of Defense: A Guide for Clinicians and
Researchers*. Washington, DC: American Psychiatric Press.

Vaillant, G. (1993) *The Wisdom of the Ego*. Cambridge, MA: Harvard
University Press.

VanderZwaag, D. (1999) "The precautionary principle in environmental law
and policy: elusive rhetoric and first embraces," *Journal of Environmental
Law and Practice* 8: 355–75.

Venter, J., et al. (2001) "The sequence of the human genome," *Science* 291:
1304–51.

Vogeli, C., et al. (2007) "Multiple chronic conditions: prevalence, health
consequences, and implications for quality, care management, and costs,"
Journal of General Internal Medicine 22(3): 391–5.

Walters, L., and Palmer, J. (1997) *The Ethics of Human Gene Therapy*. New
York: Oxford University Press.

Wang, H., et al. (2012) "Age-specific and sex-specific mortality in 187
countries, 1970–2010: a systematic analysis for the global burden of
disease study 2010," *The Lancet* 380(9859): 2071–94.

Watanabe, T., et al. (2008) "Viral RNA polymerase complex promotes
optimal growth of 1918 virus in the lower respiratory tract of ferrets,"
Proceedings of the National Academy of Sciences 106(2): 587–91.

Watman, M. (2008) "James Watson's genome sequenced at high speed," *Nature* 452: 788.

Warrier, V., et al. (2017) "Genome-wide meta-analysis of cognitive empathy: heritability, and correlates with sex, neuropsychiatric conditions and cognition," *Molecular Psychiatry*, doi.org/10.1038/mp.2017.122.

Weinhold, B. (2006) "Epigenetics and the science of change," *Environmental Health Perspectives* 114(3): A160–A167.

Weiser, M., et al. (2008) "Advanced parental age at birth is associated with poorer social functioning in adolescent males: shedding light on a core symptom of schizophrenia and autism," *Schizophrenia Bulletin* 34(6): 1042–6.

Weon, B. M., and Je, J. H. (2009) "Theoretical estimation of maximum human lifespan," *Biogerontology* 10(1): 65–71.

Wikler, D. (1999) "Can we learn from eugenics?" *Journal of Medical Ethics* 25(2): 183–94.

Wilkinson, S. (2008) "'Eugenics talk' and the language of bioethics," *Journal of Medical Ethics* 34(6): 467–71.

Williams, A. (1997) "Intergenerational equity: an exploration of the 'fair innings' argument," *Health Economics* 6(2): 117–32.

Wilson, T. D. (2011) *Redirect: The Surprising New Science of Psychological Change*. New York: Little, Brown.

Wooton, D. (ed.) (2008) *Modern Political Thought: Readings from Machiavelli to Nietzsche*. 2nd edn, Indianapolis: Hackett.

World Health Organization (2014) *Quantitative Risk Assessment of the Effects of Climate Change on Selected Causes of Death, 2030s and 2050s*. Geneva: World Health Organization.

Yehuda, R., et al. (2016) "Holocaust exposure induced intergenerational effects on FKBP5 methylation," *Biological Psychiatry* 80(5): 372–80.

Yu, Y., et al. (2009) "Evaluation of blastomere biopsy using a mouse model indicates the potential high risk of neurodegenerative disorders in the offspring," *Molecular & Cellular Proteomics* 8(7): 1490–500.

Zagzebski, L. (1996) *Virtues of the Mind: An Inquiry into the Nature of Virtue and the Ethical Foundations of Knowledge*. Cambridge: Cambridge University Press.

Index